PETERSEN'S BOOK OF MAN IN SPACE

THE POWER & THE GLORY

Volume Three

MAN IN SPACE VOL. 3 THE POWER & THE GLORY

Edited by Al Hall and the General Subjects editors of Specialty Publications Division.

Printed in U.S.A.

ACKNOWLEDGEMENTS

The editors of Petersen's MAN IN SPACE are indebted to the following agencies, companies and individuals for their enthusiastic cooperation in helping locate and make available the numerous photos and hundreds of bits of information which add immeasurably to the scope and content of this volume. Our profuse thanks to: the National Aeronautics and Space Administration (which supplied all photos unless otherwise credited), and especially Les Gaver, Margaret Ware and the Audio-Visual Branch, NASA Headquarters; Jack Stewart, Karla Scott, Donna Green, and NASA's Ames Research Center; John MacLeish, Jack Riley, Jim McBarron, and NASA's Johnson Space Center; Joe Jones and NASA's Marshall Space Flight Center; Barbara Rogers, Dorothy Marine, and NASA's Kennedy Space Center; Ed Russell, Jr., and IBM's Federal Systems Division; Raymond A. Deffry, C.W. Birnbaum and McDonnell Douglas; James D. Grafton and Boeing Aerospace Co.; P.G. Smith and Aero Spacelines; Don Bane and TRW's Space Systems Group; Harry Lowenkron, Laura Melim and Sovfoto; Park H. Irvine and Northrop; Mike Gentry and Technicolor, Inc.; Rocketdyne's E.F. Hogan and, Joyce Lincoln, and Anthony J. Longo, and Rockwell International Corp.; Richard Dunne and the Grumman Corp.; Richard Coe and the International Latex Corp.; Walter Braun and Paillard, Inc.; Joseph E. Yantis and the Whirlpool Corp.; and Ctein.

COVER

Liftoff! Launch of the Apollo/Saturn V from Complex 39 at Cape Kennedy. The 363-foot vehicle will send astronauts to the moon. NASA photo. Cover design by Pat Taketa.

INSIDE FRONT COVER

Nations around the world commemorate the efforts of the United States and the Soviet Union in space exploration with postage stamps. (From the collection of Kurt Raymond Hall)

Library of Congress Catalog Card Number 74-82254 ISBN 0-8227-0074-3

CONTENTS

INTRODUCTION

Volume One dealt with the beginnings of space exploration and Project Mercury. Volume Two chronicled Project Gemini (roughly the time from 1963 through 1966). This volume, third in the series on MAN IN SPACE, represents the half-way mark in Petersen Publishing Company's five-volume chronology. It covers the early history of Project Apollo through the unmanned mission of Apollo 6. However, from a standpoint of time, it overlaps the two previous volumes because Project Apollo was born even before the first American had been launched into space. Apollo spans the administrations of four United States Presidents: Eisenhower, Kennedy, Johnson and Nixon. The particular facet of it chronicled in this volume spans the first three of those four administrations.

EISENHOWER YEARS

Comments in the introduction to Volume Two may have left the impression that Eisenhower's administration had no interest in space exploration. Such is not quite the case. It was President Eisenhower who, in July 1955, announced that the U.S. contribution to the International Geophysical Year (July 1, 1957 through December 1958) would be the placing of a 20-pound scientific satellite in orbit around the earth. The idea had been proposed by, among others, the American Rocket Society years earlier. There were many problems confronting President Eisenhower at the time, thus the IGY satellite project was given a very low priority. Had this not been the case, it is very likely that America's Vanguard would have been the first artificial satellite in orbit instead of Russia's Sputnik.

NACA TO NASA

Eisenhower's administration was also responsible for the creation of the National Aeronautics and Space Administration (NASA). This was done in a move to counter the proposal by Congress that America's space program by handled by the Atomic Energy Commission. The January 1958 bill naming the AEC was sponsored by Senator Clinton Anderson and was merely one of many pieces of proposed legislation attempting to solve the problem created by Sputnik. Eisenhower opposed the bill, but was then obliged to come up with an alternate solution. Fortunately, just two days after Russia orbited Sputnik II (November 5, 1957) Eisenhower had announced the appointment of Massachusetts Institute of Technology president, James Killian, to the position of Special Assistant to the President for Science and Technology and the reconstitution of the President's Scientific Advisory Committee. It was to them he turned for his alternative. Their suggestions, drafted by President Eisenhower and presented to Congress on March 5, 1958, called for a clear definition of the nation's policy regarding space; the creation of a civilian agency to implement such a policy; and accomplishing all of this with a minimum of lost time and wasted effort. Congress acted upon the first two by authoring the National Aeronautics and Space Act of 1958, and creating the National Aeronautics and Space Administration.

To cope with the problem of lost time and wasted effort, it was decided that an old-line federal agency (one already very involved in the scientific study of space) would be the nucleus of NASA. This was the National Advisory Committee for Aeronautics (NACA). Founded in 1915 because of America's lack of progress in the field of aviation in comparison with European countries, the original 12-member panel whose task was to "supervise and direct the scientific study of the problems of flight, with a view to their practical solution . . . " had grown from $5000-per-year budget and one civil service clerk in 1915 to a 1958 budget of over 100 million dollars and 8000 civil service employees with headquarters in Washington D.C.; three laboratories (Langley, Lewis and Ames); two research stations (Wallops Island, Virginia and Edwards AFB, California), plus field offices in Dayton, Ohio and Los Angeles, California. The changeover took place October 1, 1958 when the 8000 civil service employees, who had gone home from work at NACA the day before, returned to their same jobs at the same facilities as employees of the National Aeronautics and Space Administration. With congressional approval, President Eisenhower selected T. Keith Glennan as NASA's administrator, with former NACA head Hugh Dryden as Deputy Admininstrator and Richard Horner as Associate Administrator.

Gradually, over the remaining two years of the Eisenhower administration, the fragmented programs relating to space were transferred to NASA. In addition to the programs, many research centers and facilities were likewise transferred. The administration also provided ample funding right through 1961, but the prority held by space programs was hard to catagorize during the "Eisenhower years."

THE KENNEDY ERA

Not so with Kennedy's administration. Senator John F. Kennedy and his running mate in the 1960 presidential election, Senator Lyndon Johnson, used "America's second place in space" as one of the prime shortcomings of Eisenhower's administration. After being sworn into office in January 1961, President Kennedy and Vice-President Johnson marshalled the nation for a march toward the "new frontier." They had barely begun moving toward the earth-bound objectives when, on April 12th, Russia claimed one new frontier (space) by orbiting Yuri Gagarin. Eisenhower had Sputnik, Kennedy got Vostok! Public reaction to both events was nearly the same.

However, Kennedy's reaction was quite different than that of his predecessor's. In March, Kennedy had asked NASA to review their Eisenhower budget, thus opening the door for a request for more funds. NASA responded with a request for an additional 28 percent. The Bureau of the Budget promptly trimmed the request by 60 percent, disallowing the entire sum requested for Project Apollo and the manned space-flight project to follow Project Mercury. So when Gagarin's flight opened the can of worms one more time, it didn't take long to find out about NASA's request and near-rejection. Kennedy's response was to request a budget amendment for NASA that was four times as large as the March amendment, with the greatest increase earmarked for Project Apollo. Kennedy wanted NASA's timetable compressed drastically from what it was during the Eisenhower era. Apollo was not the only project to benefit from President Kennedy's request. He wanted to see a massed attack on all facets simultaneously. Under Kennedy, NASA received the support it needed to make the United States the leader in space. It was also under Kennedy that it received new leadership: Glennan was replaced by James E. Webb as administrator. The other change was to appoint Robert Seamans, Jr., as Associate Administrator.

LBJ'S CONTRIBUTION

Kennedy's death brought Lyndon Johnson to the presidency. A successful 1964 presidential campaign kept him there. His influence on the space program actually began during the Eisenhower administration, however. It was Johnson who, as Senate Minority Leader, was named chairman of the Special Committee on Space and Astronautics in February of 1958. Three months later, this committee passed the National Aeronautics and Space Act of 1958. It was also Johnson who is generally credited with advancing the idea of a Space Council on which he later served. He was continually active in his efforts to keep both civilian and military emphasis on space at its peak. As the most powerful and influential member of the Senate, Johnson strove constantly in the interests of a coordinated space effort. During his tenure as America's Vice-President, he was officially acknowledged by JFK for his expertise on the subject. In fact, when the Gagarin flight shook Washington, it was to Lyndon Johnson that President Kennedy turned for an answer as to what space projects should be given the top priorities.

When Lyndon B. Johnson shouldered the burdens of the nation's highest office under less than desirable circumstances, he continued the efforts to put America first in space. He unstintingly supported NASA through the embryonic stages of Project Apollo, because he believed firmly in man's place in space. His election victory in 1964 virtually assured the nation of a dominant place in space and brought to mind the words spoken by his predecessor in that historic address of May 25, 1961: "For while we cannot guarantee that we shall one day be first, we can guarantee that any failure to make this effort will make us last." The "great society" Lyndon Johnson championed would not tolerate being last. It was during Johnson's administration that NASA became stabilized and went on to become one of the nation's largest and most powerful federal agencies. It was a fitting tribute that after his passing in 1973, the NASA facility in his home state of Texas was renamed the "Lyndon B. Johnson Space Center" (formerly the Manned Spacecraft Center in Houston).

MAN GETS A STARRING ROLE

NASA's responsibilities were broadly defined by the National Aeronautics and Space Act of 1958. Section 102, Paragraph 'A' of the Act entrusts NASA with conducting the space activities of the United States for peaceful purposes to the benefit of all mankind. Paragraph 'B' defines the civil agency's (NASA's) duties in the areas of space and aeronautics, reserving those facets relating to weapons systems, military operations and national defense for control by the Department of Defense. This, then, establishes a separation between civilian and military responsibilities. In paragraph 'C' are the eight objectives that Congress established for NASA. The fifth objective is cited as the preservation of the United States' role of leadership in aeronautical and space science and technology. In 1958, many felt the word "preservation" was ludicrous and that "restoration" was closer to the truth.

None of the eight objectives is specific on the subject of "man" in space. And, although the objectives provided for long range efforts, short-range goals and men in space did not figure prominently in NASA's directives. Vostok, Yuri Gagarin and President Kennedy's May 25th message took care of those two problems, but created a multitude of others. Getting man in space suddenly became NASA's immediate challenge and getting him to the moon and back became its goal.

This goal of a manned lunar landing soon became universally known as: Project Apollo! Its awesomeness stirred the imaginations of the populace. A London bookmaker gave 1000-to-1 odds against the United States accomplishing the goal, "...before this decade is out," and David Threlfall, a British subject who believed that the U.S. could accomplish anything it set out to do, plunked down ten pounds and accepted the wager.

He wasn't the only one to make a bundle on America's decision to land on the moon. Toy manufacturers made a commitment by investing untold thousands of dollars on products...anticipating the fulfillment of the President's challenge. The toys and the advertising of them gave a boost to the interest in America's space program, while the space program helped popularize these toys. Most young boys of the Sixties can recall Major Matt Mason from Mattel, Inc. He was a six-inch-tall astronaut doll ($2.98) who inhabited the Moon Base Space Station (sold separately, $7.98) with top-mounted beacon and tinted, flip-up solar screens, and who chugged around the lunar surface in a battery-powered, stilt-legged Space Crawler ($5.49) that doubled as a crane when locked atop the Space Station, or a Uni-Tread Vehicle ($5.49) which towed a habitable, clear-plastic sphere known as the Space Bubble ($2.49), in the company of his three buddies, Sgt. Storm, Jeff Long and Doug Davis ($2.98 each). Mattel, Inc., manufactured a host of smaller acessories for Major Matt Mason and his buddies that ranged in price from $1.49 to $2.49. Although by no means as successful as their Barbie doll for girls, Mattel sold over one million Major Matt Mason dolls and a staggering quantity of accessories. With this doll, Mattel, Inc., performed a genuine service by giving youngsters an enjoyable introduction to lunar learning, and their parents a sample of the high cost of lunar exploration!

WHERE'S MAJOR MATT MASON?

Mattel, Inc., was merely one of many enterprising manufacturers to create their own "space program." Plastic hobby kit makers such as AMT, Airfix, Hawk, Monogram and Revell were deeply involved in duplicating in miniature virtually every major piece of hardware that NASA and its contractors created for the manned lunar landing effort. Where are these artifacts today? Broken, discarded and gone, with but a few examples jealously treasured by collectors. How quickly they are forgotten.

NASA has not forgotten the agonizing struggle they fought to come from behind, and like those collectors of artifacts just discussed, NASA has taken pains to preserve the records of America's march to the moon. In case you haven't yet glanced at the acknowledgements on page 2 of this and the other volumes, it might be well to point out that every photo in every volume of MAN IN SPACE has been supplied by NASA (unless otherwise credited). They have made it possible for MAN IN SPACE to chronicle the early days of Project Apollo so that the beginning and the end can be viewed in proper perspective. What a shame that the master builders for the Pharaohs couldn't do likewise. There never seems to be a need at the time to record the obvious. Only when time blots out past events does it seem important to have such a record. The pyramids of Egypt, the monoliths of Stonehenge and the gleaming plaque with strange inscriptions fastened to an object resting on the soil of a distant world all bear testimony to the great achievements of the human race. The complete story of the former two are beyond ken, but the latter, although growing dim, is not yet too distant.

Petersen's MAN IN SPACE series is a record of how and when and why that gleaming plaque rests where it is. Volume Three is an explanation of the foundation work behind that plaque.

You will see in this volume how NASA obtained for the United States, not just leadership, but dominance in space science and technology. Whether they "preserved" it or "restored" it is a moot point. Another moot point concerns the race for the moon. It's true that Russia never actually stated its intent to race the United States to the moon or anywhere else. But the idea of warning an adversary of one's intentions has never been popular . . . in love or war! The cardinal rule of a good street fighter is not to let an opponent know you're going to do battle with him until you put him on the ground. Words and actions have precipitated battles as often as they have avoided them. However, Sputnik and Vostok taught America it is far more costly to play "catch up" than it is to play "keep ahead."

THE EDITORS

Shortly after NASA's formation and the launch of America's first satellite, the Department of Defense transferred its contract for a million-pound-thrust engine to the civilian space agency. Proposed originally by Dr. Wernher von Braun, this concept forms the basis for the most adventurous undertaking of all time—landing a man on the moon. To accomplish this will require a change in payloads from a few pounds to many tons. Two solutions to the need for more power are relatively simple: utilising existing rocket engines in clusters, as in the Saturn 1, or attaching solid-propellant booster engines to existing launch vehicles, as in the case of the Titan III-C. These boosters allow large payloads to be placed into earth orbit, but attaining the goal of a manned lunar round trip requires something new and much larger. Project Apollo is the name given to this mighty undertaking, a name taken from the Greek and Roman god of the sun. Three interdependent Apollo spacecraft, or "modules," will go to the moon; what will take

NASA PICKS THE POWER

In a race to the moon, he who has the power gets the glory

1

them there is a monstrous new three-stage rocket called the Saturn V. NASA accelerates the developmental process greatly by utilizing existing models to evolve new components. Thus, the Saturn I utilizes propellant tanks and other components from both Redstone and Jupiter missiles; its first stage is powered by a cluster of eight engines of the Thor/Jupiter type, uprated and designated the H-1. The second stage uses an engine adapted from the Atlas-Centaur program. As the size of the payloads and objectives becomes more ambitious, so do the launch vehicles, as indicated by these scale models (**1**). From left, they are: Mercury-Redstone, Mercury-Atlas, Gemini-Titan, Saturn I, Saturn IB and Saturn V. These are followed by models (to a larger scale) of Mercury, Gemini, and Apollo spacecraft. In a similar manner, the rocket engines utilized by NASA have grown in size and complexity (**2**). All are similar in that they utilize a liquid fuel, which burns when ignited with an oxidizer, usually liquid oxygen. The use of liquified hydrogen as a fuel in conjunction with liquid oxygen is fairly new, and the properties of liquid hydrogen under weightless conditions is the subject of considerable research. NASA has used the famed X-15, built by North American and powered by a Thiokol rocket engine, for research in high-altitude operation of rockets, and more recently (**3**) to test ablative heat shields by coating the aircraft with the material. Here (**4**) a Saturn 1B model is prepared for wind-tunnel testing in a simulated launch.

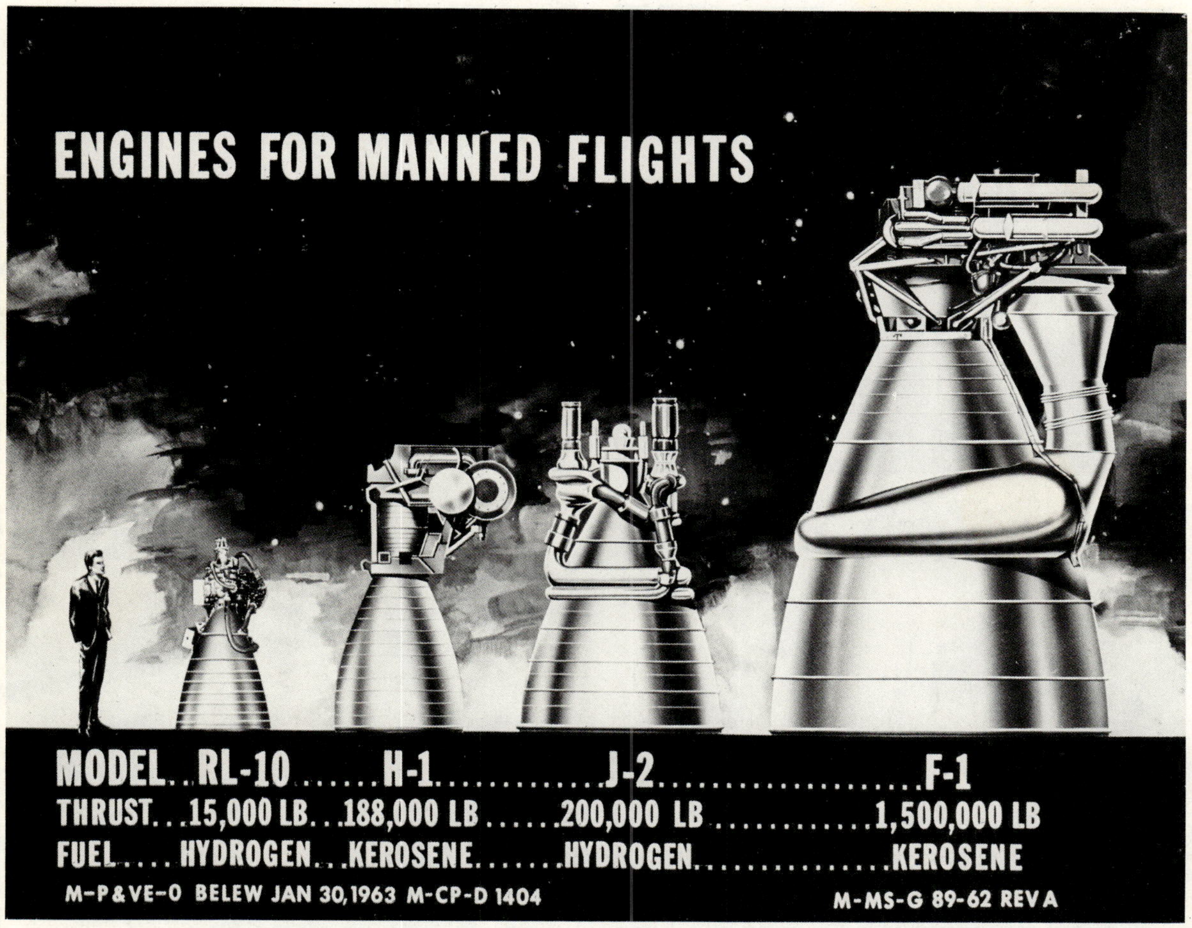

4

NASA PICKS THE POWER

This is Rocketdyne's masterpiece (**1**) and world's most powerful rocket engine, the F-1. It generates 1,522,000 pounds of thrust, or over 750 tons of push. Developed by the Canoga Park, California-based company, a division of North American Aviation, Inc., under contract to NASA, five such engines are needed to lift the Saturn/Apollo spacecraft off the launch pad. The F-1's specifications are staggering. It is 19.2 feet overall, with a 12.5-foot-diameter nozzle; its weight is 18,600 pounds dry. There is more to producing thrust than making the engine large. With an increase in size and thrust comes a tremendous increase in appetite. A reliable fuel and oxidizer pump is one of the most critical design factors. The turbine-driven pump, upper left, is only five feet in length, four feet in diameter and weighs about 3150 pounds. At operating speed (5492 rpm) this Mark 10 pump delivers 25,063 gallons per minute of liquid oxygen and 15,620 gallons per minute of RP-1 fuel. The turbine, on a common shaft with both

1

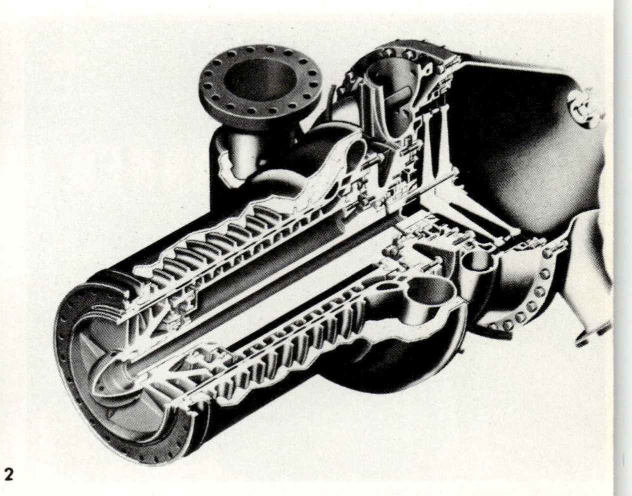

2

4

fuel and oxidizer pumps, is rated at 53,146 brake horsepower! The turbine is driven by oxidizer which has passed through a jacket in the engine's nozzle. Part of the oxidizer is passed off into the scroll manifold seen surrounding the engine skirt. Although it's only about two and one half times the size of the earlier RL-10 engine, the F-1 delivers about 100 times the thrust. And this is utilizing what is basically refined kerosene as fuel rather than high-powered liquid hydrogen. While the F-1 may be the superstar of Apollo, the J-2 is the workhorse. Also a Rocketdyne product, J-2s are shown (**4**) under assembly in Canoga Park. The J-2 powers the S-IIC Saturn V second stage in a cluster of five, while a single J-2 is fitted to the S-IVB third stage. The J-2 is 133 inches long, 80.5 inches in diameter and produces 230,000 pounds of thrust for its 3,492-pound weight. Two turbopumps deliver fuel and liquid oxygen oxidizer to the J-2. The Mark 15-F seven-stage fuel turbopump is seen at (**2**). Modifications to the original J-2 have given it a restart capability, needed for the translunar burn. This cutaway of the Mark 3 turbopump for the H-1 engine (**3**) indicates some of the materials which have made feasible some of the technical advances in the space program. Although large, these stainless steel bands (**5**) must vary no more than .006-inch from a perfect circle throughout their diameter. They fit around the nozzles of rocket engines. Thrust chambers for the F-1 engine are brazed in this huge gas (**6**) furnace under 2300-degree F temperatures.

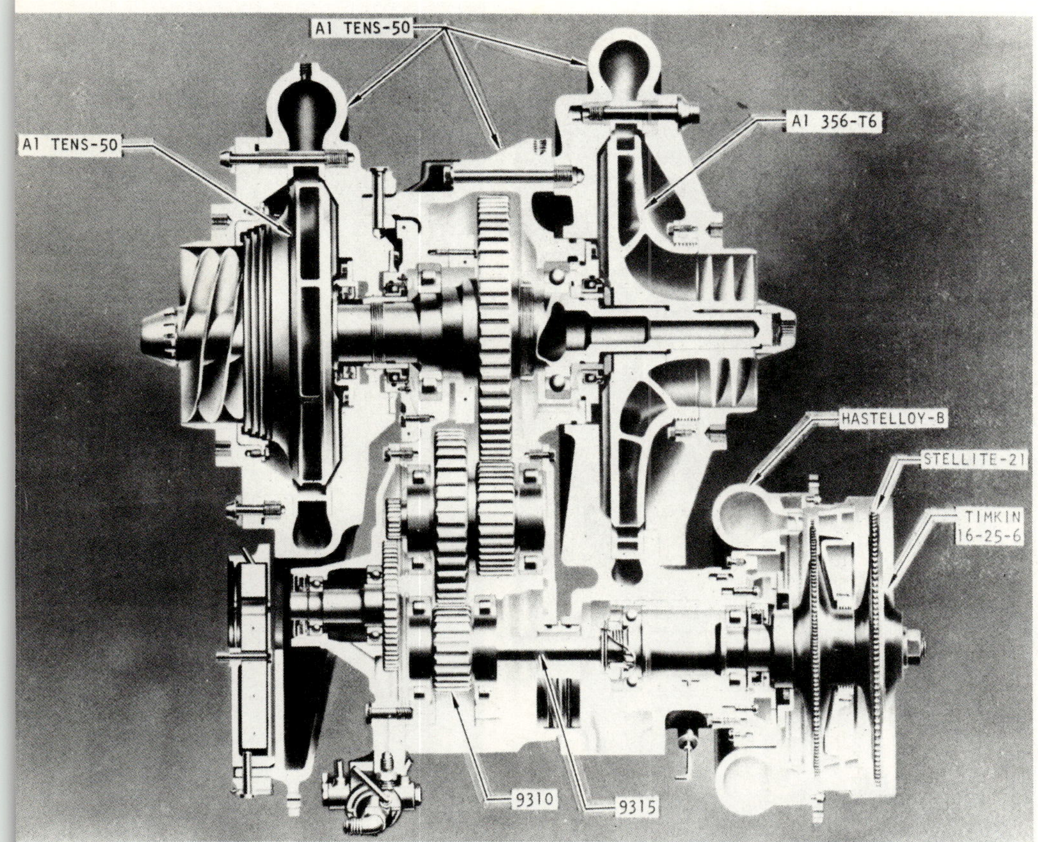

ALL PHOTOS COURTESY ROCKWELL INTERNATIONAL

6

NASA PICKS THE POWER

The skills of America's craftsmen transform the nation's industrial prowess into millions upon millions of components for the launch vehicles, the symbols of the space age. Unlike the sort from which they evolved, the Apollo generation of launch vehicles and the engines which drive them toward the stars are developed for peaceful purposes; their cargoes are not warheads, but men, not destruction but minds dedicated to scientific exploration and technological expansion. There is ample cause for awe: the J-2 engines mounted to this (**1**) second stage of Saturn V dwarf the man peering into its depths, seemingly transfixed as he might be before such a seat of harnessed power. Yet it is not so complex as he, or the patient wizardry of thousands like him who together made it; no alchemist's brew was ever so potent as this creation capable of taking common gases of the air and distilling them into raw power. At Rocketdyne's test facilities Edwards Air Force Base (**2**) each F-1 built thunders to life before delivery. Though

1

2

4

5

the F-1 is designed to burn continually through only one run, the J-2 is cycled through a start-restart to test its ability to be fired twice; once to place the Apollo spacecraft in orbit, the second to start it on its way to the moon. With every new engine, and its parts, there must be new ways devised to test it—and tests to cross-check the tests. There is a kind of beauty to the inadvertant industrial sculpture these procedures sometimes create, such as the water flow test of an F-1 (**3**). This check ensures that the hundreds of tubes, through which the oxidizer passes, have no breaks or blockages. The oxidizer, passing around the nozzle, cools it during engine operation. In order to double check the effects of super-cold fuels, both high-strength metals and delicate sub-systems are plunged through low-temperature tests, and come out looking as though they'd been through a Siberian winter (**4**). But all the engines aren't giants. Here (**6**) one of the attitude control engines, which position the Apollo spacecraft during terrestrial re-entry, has its weight calibrated to within one one-hundredth of a pound by technicians whose protective garb minimizes the chance of dust upsetting their delicate measurements. But though no part is all-important when each is necessary, it is to these which shake our earth that the eye is inevitably drawn. The H-1 engines in this view (**5**) of the first stage of the 1B first stage aptly demonstrate the cluster concept. The metal flanges which flare out protect the outer engines from aerodynamic drag.

3

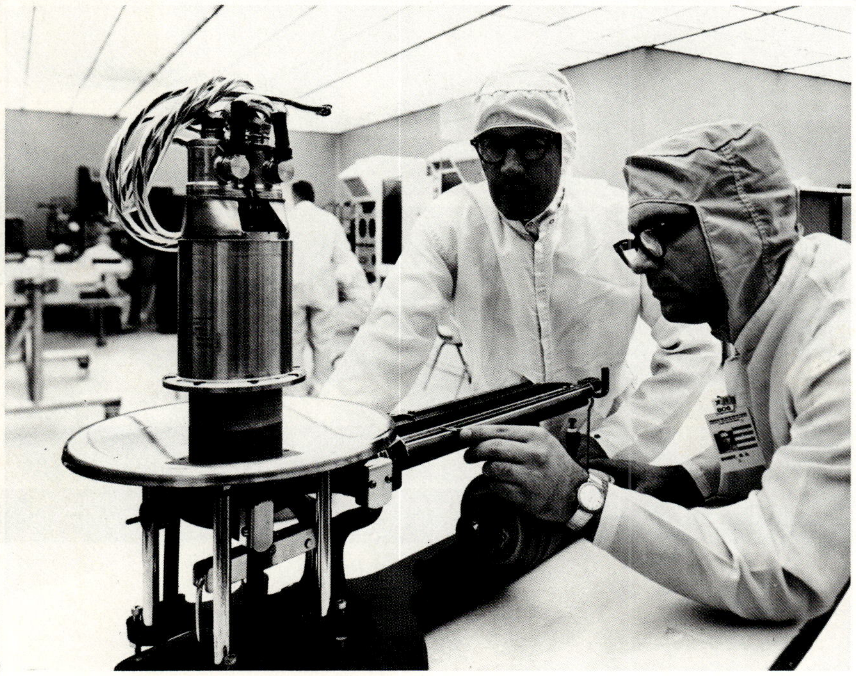

PROJECT APOLLO

Its goal is to land a man on the moon and return him safely

NASA formally initiated lunar-landing studies in 1960, before the first American went into orbit. As a result of these studies, President John F. Kennedy, in a speech given May 25, 1961, issues the challenge: "*. . . I believe that this nation should commit itself to achieving the goal, before this decade is out, of landing a man on the moon and returning him safely to earth. No single space project in this period will be more impressive to mankind, or more important for the long-range exploration of space; and none will be so difficult or expensive to accomplish.*" Funds for the development of the Apollo space craft are allocated and Project Apollo officially gets under way late in 1961. Its goal is a lunar landing, and the project is to be accomplished in three phases over a period of eight to ten years. Various groups studying the project submit plans for two alternative designs. The first, Nova Earth Orbit Rendezvous, suggests the use of two Saturn C-5 rockets (later called the Saturn V) and an earth orbit rendezvous to mate

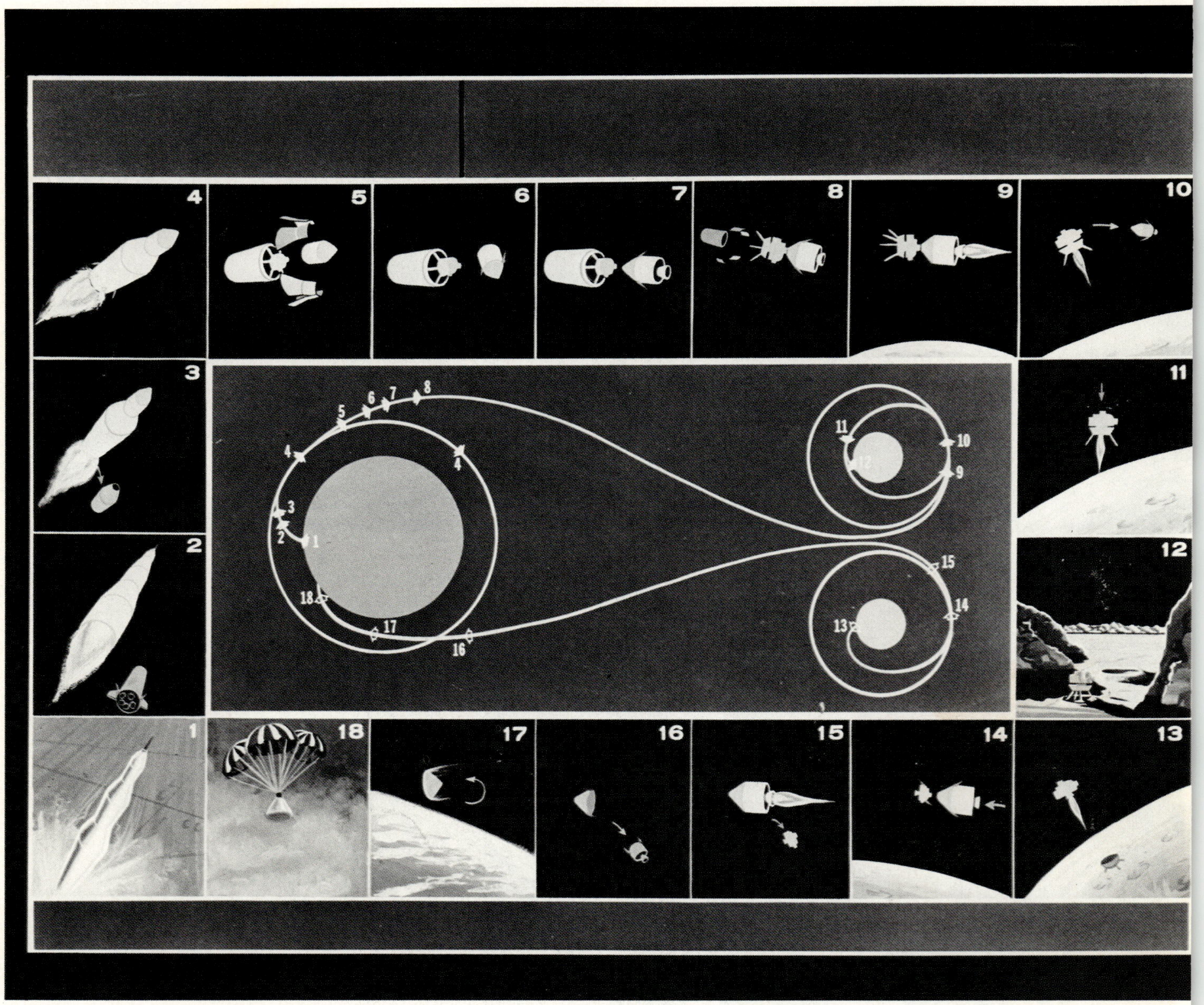

1

the spacecraft module plus an earth-to-moon orbit stage, allowing the entire spacecraft to land by employing a third module to slow the command and service modules to a moon landing. Service module propulsion would be used to launch the craft from the lunar surface. The alternative, Saturn C-5 Lunar Orbit Rendezvous, is a single Saturn C-5 launch vehicle carrying the entire spacecraft, consisting of three separate vehicles. Two astronauts will man the third module to the lunar surface from lunar orbit and return to the command and service modules while the third astronaut waits in lunar orbit to rendezvous with it. The single Saturn C-5 approach meets with little enthusiasm at first, but is finally accepted. The final Apollo flight profile sequence is depicted here (**1**). This drawing (**3**) is of the lunar landing module, which is important to the Lunar Orbit rendezvous technique. To leave the lunar surface, the upper half of the module will blast off, using the lower half as a launching platform. The upper half, containing the astronauts, will be propelled back into lunar orbit (**2**), where it will dock with the orbiting Command/Service Module (CSM). The Lunar Module (LM) and the CSM must match velocities in order to dock. To accomplish this, the CSM pilot will maneuver until the two spacecraft are locked together. The two astronauts will then crawl from the LM back into the CSM, and the combined spacecraft will orbit the moon once more before the LM is abandoned and astronauts head for earth aboard the CSM.

PROJECT APOLLO

Now that NASA has made its decision and selected the Saturn V Lunar Orbit Rendezvous design as the method of landing man on the moon, contracts will be let to industry and work towards achieving the goal President Kennedy thrust upon American technology will begin. NASA has the plans and the hardware, but now they must combine them to get Project Apollo off the ground and onto the moon. In addition to completing preparations relating to mission selection, specifications preparation and contractor selection for the major modules, work is also being done on the navigation and guidance systems. The Ranger, Surveyor and Lunar Orbiter missions have already provided enough information for NASA to select possible lunar landing sites, but the spacecraft cannot get there unless the navigational and guidance systems are completed concurrently with the work on the spacecraft itself. The artist's depiction (**1**) shows the three steps necessary to achieve Kennedy's challenge: The first is to place a spacecraft

1

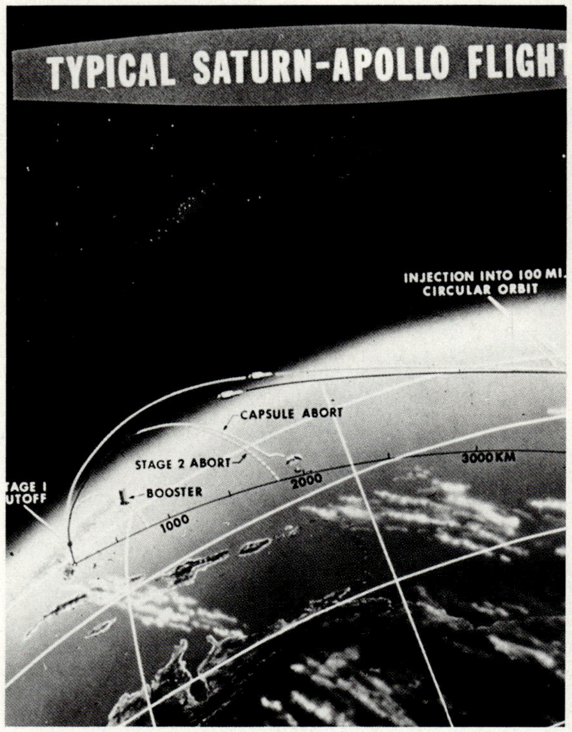

2

3

in earth orbit; the second step is an elliptical orbit around the moon—similar to earth orbit but making a giant ellipse around the earth and the moon—and finally, the ultimate goal, a manned lunar landing. This drawing of a typical Saturn/Apollo flight (**2**) gives a detailed look at the first steps for an Apollo launch. Saturn/Apollo rises off its pad like an elevator, the second stage cuts off and falls away, and the third stage J-2 engine ignites. The third stage engine burns for about two minutes and then shuts off. By that time, Apollo's velocity is more than 25,000 feet per second—and the spacecraft is in orbit. Unfortunately, there is no cameraman standing by to follow in a second spacecraft and photograph the various stages the spacecraft under goes after it leaves earth-bound camera range. An artist's concept depicting the separation of the interstage coupling ring from Apollo/Saturn V's second stage (**3**) will have to do. Separation takes place about three minutes after the spacecraft leaves the launching pad and rises into the atmosphere. The interstage ring links the Apollo/Saturn vehicle's first and second stages. Here again (**4**), we rely upon an artist's concept to show Apollo's outbound voyage. The Saturn's first and second stages have drifted away, the third stage S-IVB has finished its translunar burn and been jettisoned, and the Apollo Command/Service Module is depicted beginning its three-day translunar coast from earth. It's destination . . . the moon! And answers to some of man's questions.

4

While the basic design concepts for the Mercury capsule are being formulated in 1958, it becomes apparent that an inexpensive rocket is needed very soon to test the launch escape system and spacecraft stability. In less than a year, Little Joe is conceived, designed, and produced. It is little more than a cluster of solid rocket motors housed in an airframe, with fins for stability. Since the mission requires only powered flight in a ballistic trajectory, the flight path is simply determined by the angle at which the launcher is set (**4**). No electronic guidance is used except in the last two Little Joe II/Apollo flights. All this adds up to a vehicle that can develop a maximum of 250,000 pounds of thrust, making it more powerful than a Redstone rocket, yet costing only one-fifth as much. The ground support equipment is equally simple and inexpensive (**2,4**). From 1959 to 1961, flight tests of the Mercury capsule (Little Joe I) are conducted at Wallops Island, Virginia (**1**). The Army's White Sands Missile Range in New Mexico is

LITTLE JOE

The space program gets a solid boost from this unheralded workhorse

1

2

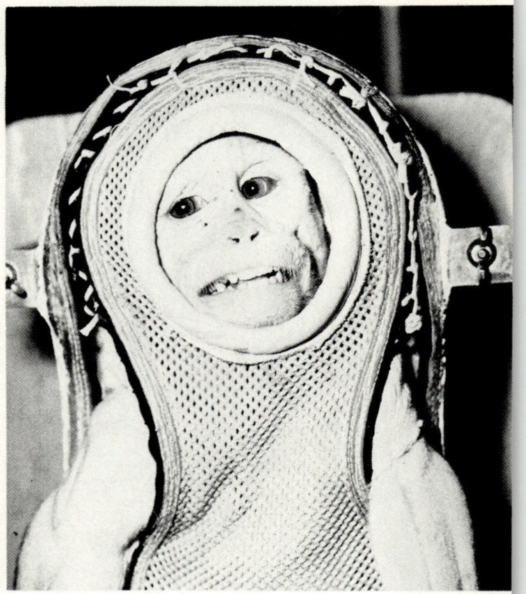

3

the site for the later Apollo tests (**6**). Performance at the point of greatest aerodynamic stress during powered flight, or "max q" in engineering parlance, is the main focus of concern. If vehicle wobble or vibration isn't too great at that time, and if the escape system performs adequately, then the design is sound. The emergency escape system is a rocket mounted on a pylon attached to the top of the capsule (**5**). If something goes wrong, explosive charges separate the capsule from the launch vehicle. At the same time, the escape rocket is fired, pulling the capsule up and away; parachutes are then deployed. So on December 4, 1959, a rhesus monkey named Sam (**3**) takes a ride. When Little Joe burns out, the escape rocket pulls Sam to 280,000 feet. Weightless for three and one-half minutes, the little fellow comes through re-entry and impact in good shape. Equally significant is that the capsule escape system functioned perfectly at very high speeds. Six weeks later a female monkey named Miss Sam is sent aloft. This time the emergency escape system fires at an altitude of nine miles, right in the region of max q. Both during ascent and descent, Miss Sam performs certain tasks well; however, she fails to respond for about 30 seconds during the abort. This, plus the high measured noise level inside the capsule, causes some concern about astronaut comfort and effectiveness. But since hardware performance at max q is proven with this boilerplate capsule, the next goal is to do the same with a regular production model.

4

5

6

LITTLE JOE

On April 28, 1961 Little Joe again takes off. One motor fires five seconds late, so the vehicle climbs to only 14,600 feet instead of 45,000 feet. But because it also travels at a much lower angle, the dynamic pressure is about twice the normal maximum. The abort is on schedule and everything else happens as it should. Therefore the flight more than proves the structural integrity and escape system performance. The Little Joe I program thus ends successfully, despite the odd last flight path. The capsule escape system is now "man-rated". With Little Joe II and Apollo capsules, the main concern is again with performance at max q. As with the former flight series, verification of the performance of the Northrop-built parachutes is an additional test objective. In (**1**) we see a closeup of the parachute compartment. Parachute deployment in normal descent is shown in (**3**). Because an abort may occur anywhere, a safe, soft landing must be made on land (**4**) as well as water. The first Apollo spacecraft test flight is made on

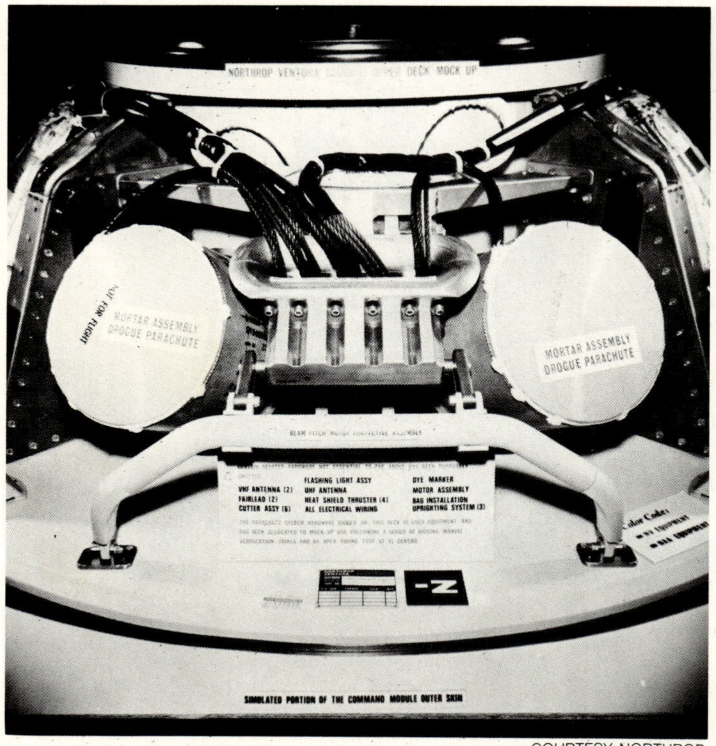

1

COURTESY NORTHROP

2

4

May 13, 1964, with a boilerplate Command and Service Module (**2**). When max q occurs at 17,000 feet, Little Joe is intentionally exploded. The escape rocket (**5**) functions perfectly, but excessive oscillation on the way down fouls the lines of one parachute. Nevertheless a safe descent is made with only two of the three Northrop parachutes, and a comfortable safety factor is demonstrated. The second flight, on December 8, is another max q abort test with a boilerplate spacecraft. This is, however, the first test for the Command Module's cover. This cover, to be jettisoned on a real Apollo flight after orbit is achieved, has two functions; it protects the Lunar Module docking mechanism on top of the Command Module, and it protects the spacecraft windows from the exhaust of the launch escape motor. Again, this flight is highly successful. The story is different for the May 19 launch, however. Soon after liftoff, a guidance system malfunction causes such excessive roll that the vehicle comes apart. The escape system performs perfectly and the capsule is recovered. Strictly speaking, the test flight objectives were not met but the system does prove itself in an actual emergency. The last flight on January 20,1966, is to qualify a real Command Module. Again, the abort at max q is successful. The canards, or little wings, on the escape rocket also show their ability to turn the spacecraft around after an abort. With the Apollo escape system now also fully "man-rated", the job of Little Joe, stubby, ungainly, but effective, is done.

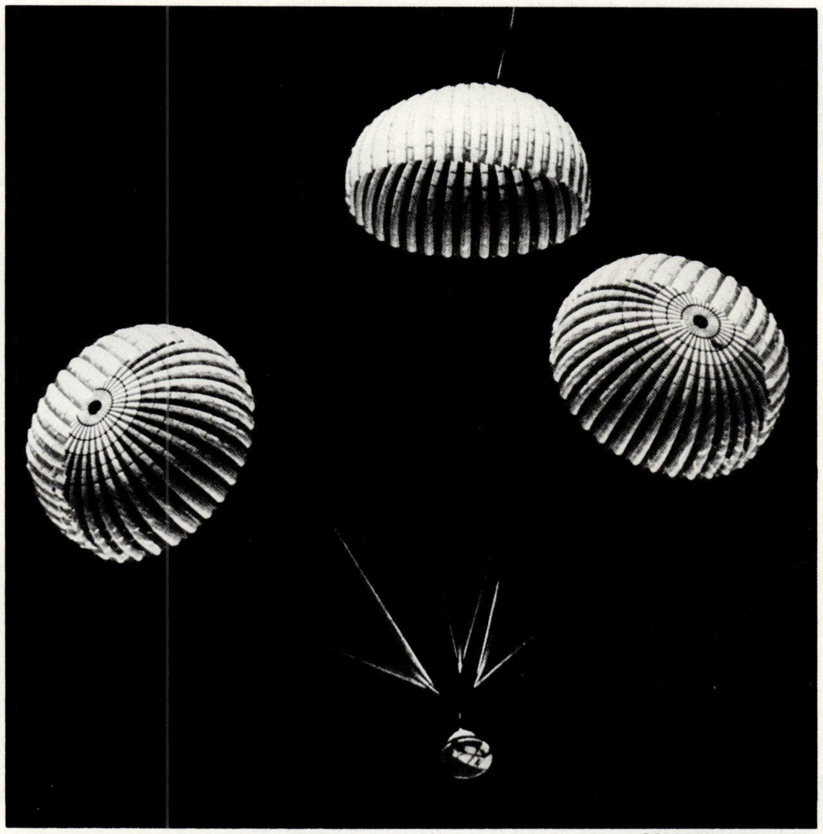

3

COURTESY NORTHROP

5

Space commitment: on May 25, 1961, President John F. Kennedy stresses the importance of landing a U.S. astronaut on the moon within the decade and bringing him back safely. Coincidentally, only two weeks later, Wernher von Braun bluntly states that America already has an instrument capable of achieving the goal. The "instrument" von Braun refers to is the powerful Saturn One launch vehicle. In dramatic elephant/mouse comparison, the awesome Saturn One (**1**) dwarfs the Redstone rocket used to launch Project Mercury. The Saturn One's statistics are impressive. It stands a towering 162 feet above ground—equivalent to a 16-story building. There are three powered stages to it, plus payload. The base, S-I booster stage, consists of eight H-1 clustered engines, producing 1.5 million pounds of thrust. The S-IV second stage houses four engines using liquid hydrogen and liquid oxygen propellants. A total of 80,000 pounds of thrust is produced by the second stage. The S-V third stage utilizes two engines of the LR-115

THE SUCCESSFUL SATURN 1

America's most powerful rocket launches Apollo on a note of success

1

2

3

type used in the S-IV stage, producing 40,000 pounds of thrust. Later, NASA approves an increase to six LR-115 engines in the S-IV stage. The diameters of stages S-I and S-IV are 257 and 220 inches, respectively, each diameter being roughly equivalent to the length of a Cadillac. The S-V third stage measures 120 inches across. At Cape Canaveral, Pad 34 (**2**) was chosen as launch site for the first test of the Saturn One. In all, NASA plans a series of 10 flights for the Saturn One program. Terminology-wise, NASA designated the initial launch as Saturn Apollo 1 (SA-1), which is to be followed by test flight number two, SA-2, at a later date. Time: October 27, 1961. A perfect day, a perfect launch! The Saturn SA-1 first stage booster (**3**) shakes the earth with 1.5 million pounds of thrust. The 925,000-pound launch vehicle, the largest ever to be tested in the world, carries water-filled dummy second and third stages. Speeding to an altitude of 84.8 miles and arcing 214.7 miles down the Atlantic Missile Range, the first of the Saturn One launches, SA-1, is deemed a success. But all is not 100 percent perfect: slight fuel sloshing is encountered near the end of the flight. The time is now April 25, 1962. At Cape Canaveral, the Saturn One SA-2 launch (**4**) is being readied. The SA-2 races skyward to an altitude of 65 miles, where the vehicle is blown up intentionally (**5**) to dump water ballast from the dummy second and third stages into the upper atmosphere; this is "Project Highwater," a high-altitude study of gas dispersion.

4

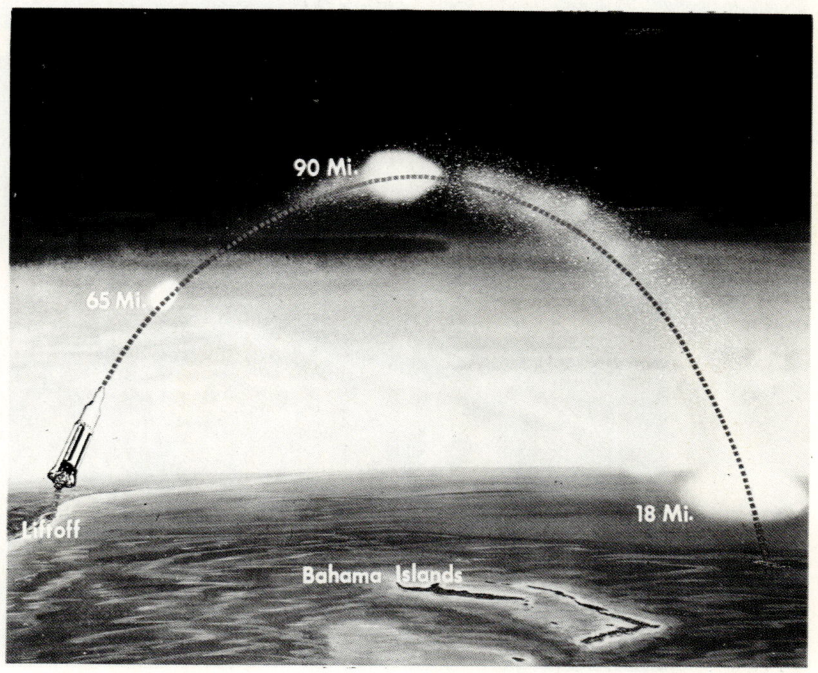

5

SATURN 1

In the recent Saturn One SA-2 flight, the fuel tank modifications were successfully done to the vehicle to eliminate the fuel sloshing problem encountered in the initial SA-1 flight. As for Project Highwater—the dispersing of water ballast into the upper atmosphere to produce massive ice clouds and lightning-like effects—it, too, went off without a hitch. But a more demanding Project Highwater II looms. Here at Cape Canaveral, launch control (**1**) monitors the preflight progress of Saturn One mission SA-3. Launch Control's voice booms, "5-4-3-2-1-0!" On November 16, 1962, Saturn flight SA-3 rises off the pad (**2**). The potent first stage booster propells the vehicle to an altitude of 104 miles, where, 4.53 minutes after lift-off, the water-filled second and third stages are detonated as in Project Highwater I. But Project Highwater II requires a more complex demand: atmospheric physics are to be studied via electronic instrumentation on earth. Unfortunately, due to poor telemetry, the resulting data gathered is much less

2

1

5

than hoped for. Preparations begin immediately for the next Saturn One launch. Here, the Saturn One's (SA-4) first stage booster (**3**) is being lifted into position. Flight SA-4's main goal is to cut power to one of the eight H-1 engines in the S-I stage, and reroute the "dead" engine's fuel to the other seven "live" engines. The water-filled dummy second and third stages of SA-4 are predominant in this unusual perspective (**4**). On March 28, 1963, everything goes as planned! The SA-4 attains a maximum velocity of 3360 miles per hour and an altitude of 80 miles. This flight is the last to be powered by only the first stage. It is also the last Saturn One flight with a three stage plus payload configuration. All future Saturn One flights would utilize two stages plus payload. Ten months pass; Saturn One flight SA-5 is impending. This will be the first Saturn One launch with the vehicle in so-called "Block II" configuration: an S-I first stage, S-IV second stage and 154-inch-diameter payload stage. Whereas the three-stage plus payload Block I configuration used in the previous Saturn One flights measured 162 feet in height, the new Block II Saturn One configuration towers 22 feet higher, standing an impressive 184 feet above ground. Here (**5**), technicians pre-flight check Saturn One SA-5 on launch day. January 29, 1964: flight SA-5 accelerates dramatically off the launch pad (**6**), trailing a long spume of rocket exhaust in its wake. Note second stage and payload. Later in flight, SA-5's S-IV stage kicks in perfectly on command.

3

4

6

SATURN 1

This unique sequence of launch photos (**1** to **8**) shows the Saturn One flight SA-6 from ignition through liftoff. The creative photos were captured from a fixed camera position using a nitrogen-purged box, atop the umbilical tower. A remotely operated 35mm Nikon camera utilizing a 21mm wide angle lens was positioned so that the action format was vertical. The 35mm Nikon also incorporated a motor-driven advance set at eight frames per second shooting speed. The film used was a 250 exposure load of ECN. Note in photo (**1**) how the flame deflector shield—anchored below the launch pad's circular top cutaway—deflects the searing, straight-down blast of initial ignition. After deflection, rocket flames are visually evident in the picture coming out 90 degrees to the left and right of the launch pad's base. This flame deflector shield plays an important part in launch pad technique. Without the deflector, thrust bounce-back or crystallization of launch pad material could occur, causing an abort of the flight. In pho-

1

2

3

4

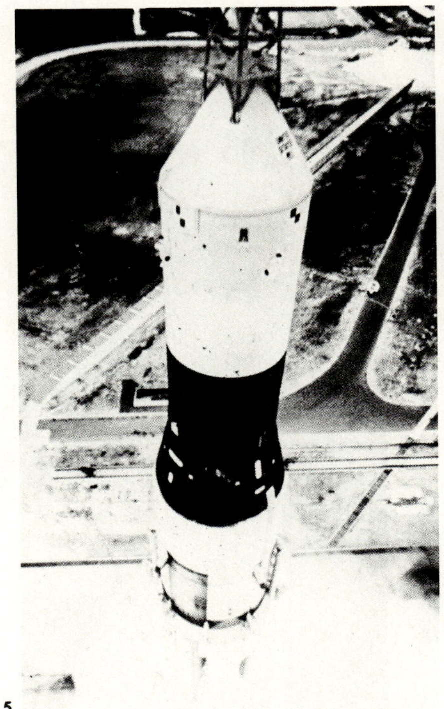

5

6

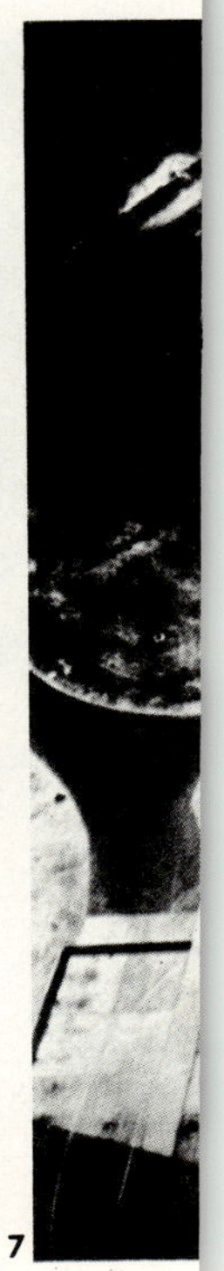

7

to (**2**), the service mast's black arm is shown connected to the vehicle. After full thrust is developed, and an instant before the Saturn 1 begins to move, the service mast disconnects itself from the vehicle (**3**) and swings unobstructively away. The little rocket (**4**) attached to the top of the Apollo boilerplate No. 13 Command and dummy Service Module is the heart of the Launch Escape System (LES). The LES will take the Command Module containing the astronauts away from the launch vehicle in case of an emergency on the pad or shortly after launch. In a deployed state, the LES carries the Command Module to a sufficient height and to the side, away from the launch vehicle, so that the earth landing sub-system can operate. As the picture clearly shows, the LES looks like a rocket connected to the Command Module by a latticework tower. Physically, the LES measures 33 feet long, weighs approximately 8000 pounds, and spans four feet across at its maximum diameter. In photo (**5**), the sheathed Apollo boilerplate No. 13 hoves into view. Simply put, boilerplates are research and development vehicles which simulate actual production spacecraft in size, shape, structure, mass, and center of gravity. SA-6 is the first flight to test the compatability of the Saturn 1 launch vehicle with a boilerplate spacecraft. Photo (**6**) focuses on the six-engined S-IV second stage, while photo (**7**) catches the eight-engined multi-chambered S-1 first stage as it passes by. And photo number (**8**)? The flames of a successful launch!

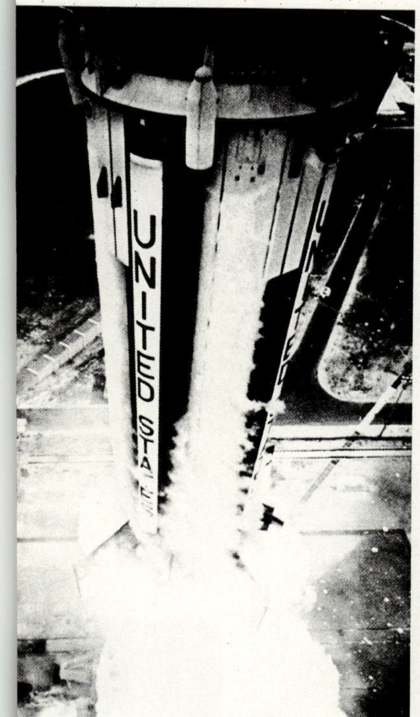

8

SATURN 1

Since the May 28, 1964, SA-6 test at Cape Kennedy, the ground technicians have been busily preparing for the next Saturn One launch—flight SA-7. SA-7 is scheduled to depart September 18, 1964. One of the herculean tasks at hand is to lift SA-7's mammoth S-I first stage off its transport dolly and into an upright position on the launch pad (**1**). After the S-IV second stage has been attached onto the first stage booster, it is time for the technicians to start mounting the Apollo boilerplate spacecraft atop the second stage (**2**). This is what flight SA-7 is primarily concerned with (**3**)—the Launch Escape Sub-system (LES). During SA-7, a high altitude abort will be performed utilizing the LES and unmanned, boilerplate Command Module. On launch day, SA-7 leaves the pad perfectly (**4**). Later, during flight, SA-7's LES functions as programmed. Here (**5**) Dr. Kurt Debus, director of NASA's John F. Kennedy Space Center, and members of the blockhouse team watch the performance plot boards—moments after launching of the

1

2

3

4

seventh consecutive successful Saturn One flight vehicle. The Saturn program continues. It is now February 16, 1965. Under cover of darkness, Saturn One flight SA-8 disappears from Cape Kennedy in a white-hot beam of light (**6**). Flight SA-8's main purpose is to orbit Pegasus I—a 3200-pound meteroid detection satellite. Through the use of extended, electrically-charged wing panels that are coupled to onboard instrumentation, the Pegasus I will be able to telemeter the frequency of meteoroid impacts upon it—back to earth. It is advantageous that NASA receive this information of meteoroids to be encountered in near-earth environment. The safety of future manned and unmanned spacecraft may beneficially be aided if accrued Pegasus data is utilized in the design stages. Operationally, the deployment of Pegasus I is ingenious but simple. Once in a stable earth orbit of approximately 300 to 800 miles, the protective housing of the payload section is jettisoned, exposing the retracted, accordian-like wing panels of Pegasus I. The meteoroid detection panels are then made to expand outwardly. With the panels fully extended and aligned in a flat plane, Pegasus I is ready for telemetering meteorid information back to earth. Dimensionally, the satellite sports an impressive wingspread of 96 feet, making it even larger than its namesake, the mythological flying horse, and has 2300 square feet of panel sensors. Orbital data: Pegasus I was successfully placed in orbit at 462 statute miles (apogee) and 308 miles (perigee). Orbital time: 97 minutes.

5

6

SATURN 1

More meteoroid detection satellites are scheduled for launching into earth orbit via Saturn Ones. May 25, 1965: Dr. Kurt Debus, Director, Kennedy Space Center, left (**1**), and Dr. Wernher von Braun, Director, Marshall Space Flight Center, watch the progress of SA-9/Pegasus II countdown at Cape Kennedy blockhouse. It's the calm before the storm (**2**). SA-9's eight H-1 first stage booster engines can produce 1.5-million pounds of thrust on demand. Note arrangement of the eight H-1 thrust nozzles. Meanwhile in the launch control room (**3**), technicians carefully monitor the countdown. Here in launch complex 37 (**4**), Dr. Wernher von Braun takes a close look at the SA-9 flight vehicle through the periscope inside the blockhouse. Booom! SA-9 (**5**) generates lift-off thrust on way to achieving its goal of orbiting Pegasus II. Note the protective, cylindrical housing of Pegasus II. The retracted, 96-foot wing of the satellite is safely stored in the cylinder area below the dark nose cone and just above the horizon line

1

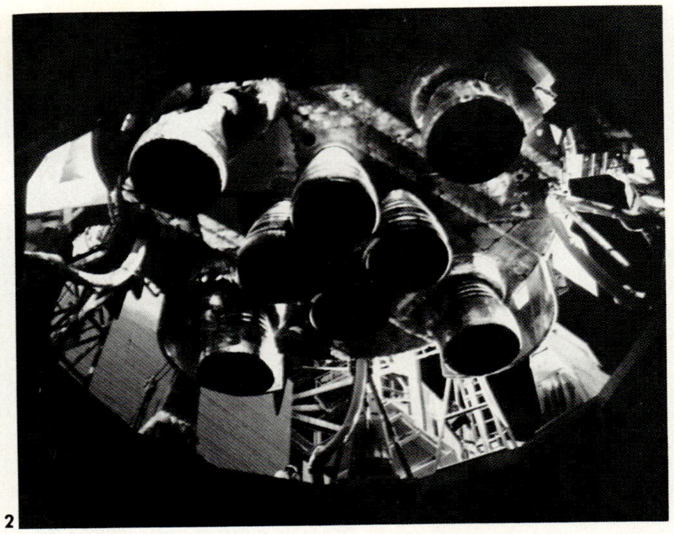

2

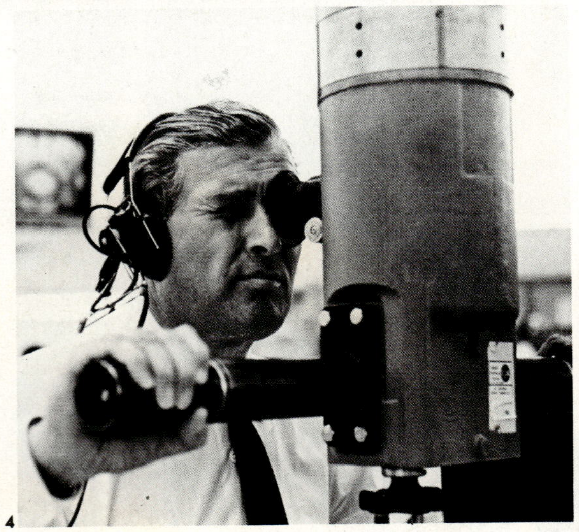

4

5

that bisects the payload section of SA-9. Minutes after launching, SA-9's goal is achieved; Pegasus II is in orbit. Orbital data: 466 statute mile apogee; 314 statute mile perigee; 97.3-minute orbiting time. Two months pass. The time is now July 30, 1965. Flight SA-10 rockets away from Cape Kennedy's launch pad (**6**) to place the third Pegasus meteoroid detection satellite into orbit. Pegasus III settles into a 336 statute mile apogee/323 statute mile perigee track, circling the earth once every 95.3 minutes. All three Pegasus meteoroid detection satellites are now simultaneously orbiting the earth, relaying valuable information. This last flight, SA-10, marks the 10th and last launch of the Saturn One series. Outstandingly, all 10 Saturn One flights have been successful—an unprecedented new success record for the U.S. space program. What lies ahead for the Saturn One launch vehicle? It will exist no more in its basic form as we have followed it. In its place, a new, more potent, uprated Saturn launch vehicle is scheduled to emerge. The new uprated Saturn will be designed to harness a completely different stage-two powerplant consisting of a single 200,000-pound-thrust liquid hydrogen/liquid oxygen engine—instead of the six 20,000-pound-thrust engines used in the basic Saturn One's second stage. Also, the uprated Saturn launch vehicle will possess a modified, more powerful first stage booster. But, enough uprated Saturn speculation; for now, let's savor the 100 percent success of the Saturn Ones!

3

6

It was a long time before the hazards to a man in space were clearly understood. By their heavy imagination, early stories merely admitted that there was simply no way to know what a man in space would have to face. Gradually, through the use of instrumented satellites and the workings of sophisticated science, it became apparent that formidable, indeed, were the hazards. A familiar environment could be built into a spacecraft, but still, the human body—separated from its mother planet by thousands of miles—would have to depend on many of its own devices to cope with the unknown. And the fundamental worry of the longer Apollo missions did not easily provide reassuring answers on some less apparent observations. For example, the Russian cosmonauts complained often of dizziness and nausea after earth-orbital flights; a malady lasting long afterwards. American astronauts spoke of "feeling a heavy weight"—much heavier than normal—when trying to sleep in their earth-bound beds following a long

TRAINING AMERICA'S ASTRONAUTS

There's more to a spaceman than being a pilot in perfect physical condition

1

2

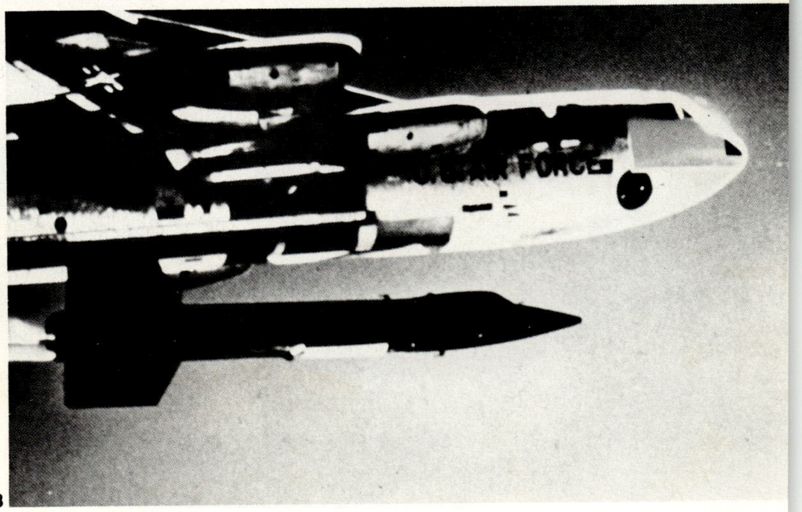

3

space flight. Post-flight complaints include bone-calcium reduction and heart palpitation. There are many, *total* answers for which science is still searching. The list of variables is long. It is natural, then, that against this background, the wings of Naval and Air Force astronauts (**1**) mean much to the men who wear them. It is a feeling known well by this group of astronauts standing on a Cape Kennedy launch pad (**2**); wondering, behind their smiles, of the training they must assimilate for the rigors they are about to face. To qualify as an astronaut, one must attain an altitude of 50 miles, though not necessarily without other training. Both X-15 pilots in the group, Neil Armstrong and Joe Engle, know that an X-15 (**3**) average maximum flight time is 9 minutes and 46 seconds. While flying Gemini 8, Armstrong could certainly tell the difference, but could feel the similar *g-loads*. An X-15 peaks its altitude curve near 250,000 feet. It re-enters the atmosphere at 3000 miles an hour. But a space capsule screams through the atmosphere back to earth as a flat rock must enter a smooth sea—with the same hazards of proper aiming—yet at speeds of up to 25,000 miles an hour! The capsule's small size is clearly shown (this one an Apollo) besides astronauts (from left) Engle, Brand and Kerwin (**4**). First American into space, Alan Shepard (**5**) withstood *16-g's* on the ALFA centrifuge trainer. What counts, though, is the future; demonstrated by Rusty Schweickart (**6**), next to a Saturn 1-B. It is a forward look (**7**), to Al Worden.

4

6

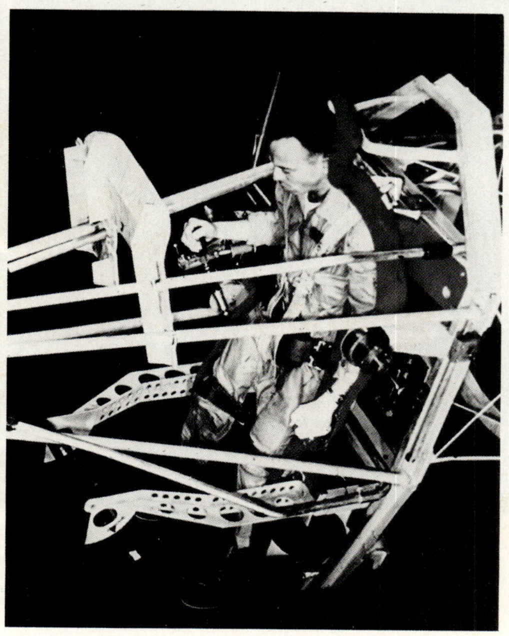

5

7

ASTRONAUTS
The men

To this point in the series, we have seen the birth of the space race—or at least a race between two types of technological sophistication. We have seen the satellites, and the successful completion of the Mercury and Gemini missions. The faces of the men who have flown them, however, might not always be with us. . .the Glenns, Coopers, the Wally Schirras, Armstrongs and "Buzz" Aldrins . . . these men—so much alike in so many ways—could easily, in our minds, become as the trade winds, passing before us and then passing on; once as individuals in the spotlight; then blurred, as a single image with the deepening shadows of time. But many NASA flight surgeons often expressed that for each of these heroes, there will be others with the same standard of qualifications and ability. Such statements are a reflection of the thoroughness of NASA's training, to say that in most respects of ability, there isn't the thickness of a piece of paper in difference between them. For each prime crew member, there will always be a back-

1

up. He is there—to back up the machinery of technology. He is the best insurance to the success of a space mission. There have been countless accounts already published on this subject—both for and against man in space. But what matters is that an extension of *every* man rides with an astronaut; not with a machine. And the point, often obscured, is that we would like to *feel* and so *understand* what it is to fly an egg-shell of metal and flashing lights . . . to really *know* what manner of a man in space will back up a failing computer at a critical moment. What sort of man can watch quietly at re-entry, as the titanium skin of his tiny craft is barbecued to a glowing, sinister, bright-red? To begin to understand is to know that, first, an astronaut has a passion to fly . . . as a test pilot, higher and faster than before. Any astronaut, however, will tell you that he is made, not born. This thought becomes a collective action, hovering in the mind of many, during a special salute at the Cape for the original seven (**1**), by now space veterans. But, in a lesser sense, an astronaut's training is many things . . . It is Buzz Aldrin (**2**) studying a map of the heavens with an interested spectator; it is NASA's acheivement of sound without echo in the anechoic chamber (**3**), an aid in calibrating the precise communication that must guide an astronaut's capsule. And, it is a continued interest in the "good earth" . . . a pause along a NASA corrider during training, by Neil Armstrong and Jim Lovell (**4**). How different, the bewitching blackness of space.

2

3

4

ASTRONAUTS
The pilots

Scott Carpenter was to write in his personal log his own words describing what "a marvelous and incredible machine the human body is." One NASA official remarked candidly that each astronaut performed and accepted the very tough training with such skill and fitness that perhaps "lots would have to be drawn to see who would be the first man in space." The original list was 500; then 110; then still more "hair-splitting," with 56 out of 69 selected as candidates. Graduating astronauts numbered 7 in 1959; 9 in 1962; 14 in 1963 and 19 in 1966. There were more Air Force pilots than Navy, but not by much. There were 6 Marine Corps pilots. The pilots whose duty was primary stateside assignment numbered more than 2 to 1 over those with combat experience. But of those who flew in World War II and Korea, all were highly decorated for skill under pressure. As a group, the average age at time of selection was 34.6; weight, 164 pounds, height, 5 feet 9½ inches; married, with 2.1 children. Most of all, each *wanted*

1

2

3

to go into space; not for personal glory (NASA found that out quickly enough) but for a deeper commitment; a thing not easy to explain. Part of that desire, though, is of the same warp and wøof that once drove him to become a test pilot; an exhilaration for flying. Nothing new; that excitement can be seen at the beginning . . . on a large scruff of asphalt about nine miles southeast of San Antonio, Texas. While most probably it began there, at Lackland Air Force Base, it could have been with the Navy, at Pensacola, or Corpus Christi. It was a memory of absorbing physical conditioning. It was straining to the mental beat of the sciences, higher mathematics, and the laws of a pilot's realm; the days at Webb Air Force Base and the Cessna T-37 twin-engine "Tweety Bird." How the memory works . . . The restrained smile of an IP (Instructor Pilot) after first solo. "Finally," into the advanced, supersonic T-38 Talon or comparable; and a feeling of sureness in the air . . . a hundred different manuevers; then—assignment to a combat wing, as a prelude to the unknown. Most astronaut candidates, alone, now, in the centrifuge (**1**), might not think of their beginnings; or have the time, in this multiexposure of the Translation and Docking Simulator (**2**)—vital to the understanding of a lunar landing. In the Apollo Mission Simulator (**3**), Borman, Collins and Anders learn its delicate feeling. A simulated Command Module control panel (**4**) helps teach instrumentation and direction of electrical impulses. And. . .the (**5**) inevitable *10-g's* gravity forces.

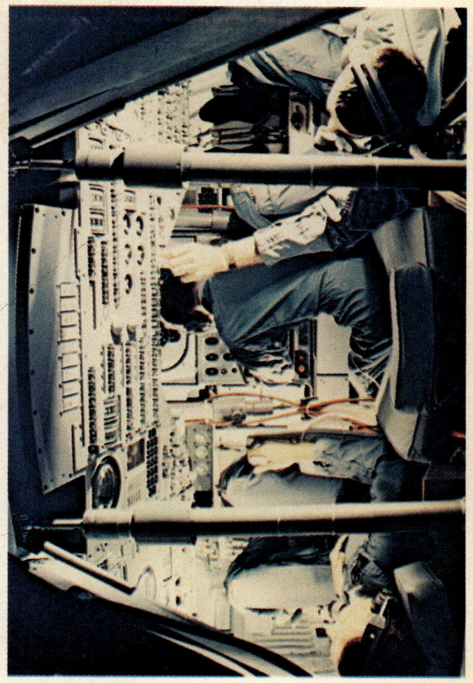

4

5

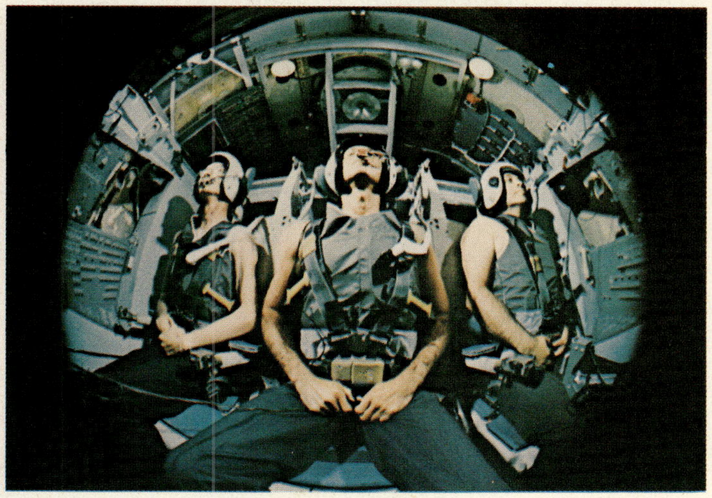

ASTRONAUTS
Survival-1

What sort of a personality is a newly graduated astronaut? In truth, he is humorous, yet serious; athletic, yet intellectual. Perhaps the voice of John Glenn is a good illustration; good because his orbit was the first; and the world has a way of remembering an American first, sometimes a little longer. John Glenn, then, becomes a convenient example of this new breed of man—the astronaut. There is nothing gaudy about Glenn. He comes from Ohio, but he could have come from anyplace. When, in one of his three orbits, his exhilaration was apparent at the sight of his world 162 miles beneath him, his words rang like a breeze through a Midwest field of wheat . . . no more, no less. "Boy," Glenn said, "the American shoreline sure looks wonderful." That is the point of the whole thing; what words cannot say. They are, these astronauts, simply one of *us,* and that's what makes us proud. There is a basic self-respect to an astronaut, that along with his humor and toughness, naturally turns jungle survival training into a "snap,"

1

2

6

5

although an astronaut will tell you, "It's best not to come down in the jungle." But in the jungles of Panama, training gives preparation, if ever a spacecraft should land in such a place. Hopefully, that will never happen. Nevertheless, finding civilization on a jungle raft is no problem (**1**), nor is finding nutrition inside a bamboo stalk to Dave Scott (**2**). Though the survival school is rugged, four of the group, plus a native guide, beam their approval (**3**). From left is Neil Armstrong, and Ed White, John Glenn and Scott Carpenter. Then it's time for dinner, and a scaly iguana gets the nod (**4**), with head chef Walt Cunningham receiving input on what he can do with the lizard from Bill Anders (right). Alternate menus include spit-roasted boa constrictor (**5**), while after-dinner tree cutting demonstrations (**6**) are also prepared by local Choco Indians. As the astronauts adapt to the Panama Jungle Survival School, it might do well to reflect on the other requirement to the training of an astronaut—the "sticky" part, made up of test scores, norms and ratings. NASA has called for a university degree in the physical sciences or engineering; a degree from one of the military test pilot training schools, and an accumulated flight time of at least 1500 hours. Most of all, astronauts must believe in space flight. (Each astronaut has 3000 to 5000 hours or more). Most of all, each astronaut must believe in the value of outer space flight. Without a shadow of a doubt, he should want to be exactly where he is; able and willing not only to endure the hazards, but to triumph.

3

4

ASTRONAUTS
Survival-2

As expressed by one NASA flight surgeon: "The only thing about the astronauts is that they're sort of super-normal." But the astronauts, cut to 32 candidates from the original 56, by a three-man NASA team (a psychologist, a test pilot and an engineer), have maintained two overpowering desires, throughout their training. One is a personal commitment to stay physically fit; the other is a satisfaction of the desire to fly. Endless hours in flight simulators have not altered the need to "be up there, like the eagles." Two F-102's have been placed at the exclusive use of the astronauts while at the Cape; and there is always an available aircraft, wherever they are; to keep the adrenalin pumping. Since they are test pilots, it is a thing they do best. Flying 30 to 50 hours a month, gives them a steadying influence to go through the endless hours of examinations for their blood, tissues, eyes, ears, nose, throat, heart, circulation, nerves, muscles; their thoughts and mental abilities, both academic and personal. Psychologically,

1

2

4

their tests have included special low-residue diets; personality evaluation, behavioral sciences, stress and fatigue limits, accelerative forces, equilibrium and vibration, high-energy noise, low barometric pressure, thermal stress, and anything else the NASA doctors could devise. The astronauts have been isolated, boiled, and frozen; spun mercilessly up to *20-g's* in the centrifuge; subjected to smashing sounds and perfect quiet—sometimes nearly simultaneously and usually as a preface for solving complex problem tests! They also have been subjected to equivalent pressures of 100 feet below the ocean's surface, and at 100,000-foot altitudes, protected only by their spacesuits They have felt weightless in specially designed and flown C-130 and KC-135 cargo aircraft. Finally, NASA judged the differences in the 32 so slight that any mission could use any astronaut; they were all interchangeable—these men with enough energy to jog in the desert's heat (**1**). But after a hypothetical spacecraft landing in the Nevada desert (**2**), such running is an absolute "no-no." There, training near the Carson Sink area, John Young maintains psychological discipline (**5**) with a battery powered shaver. Nevada was augmented by eastern Washington state, and for volcanic, "lunar-like" field geology, in the inhospitable region of Mexico's Sonora desert. Liferafts were used as mattresses; parachutes were cut into clothing and shelters (**4**); put up and torn down again (**6**), in an endless question and answer (**3**) regimen.

3

6

ASTRONAUTS
The moment

With an uncommon curiosity; and finding answers to new questions, a special kind of pride comes to a test pilot . . . and to an astronaut. It's what NASA looks for. As an example, look at Deke Slayton, who flew B-25's and B-26's for 63 combat missions in World War II. While a test pilot at Edwards Air Force Base in California, Slayton flew every type of jet fighter in the Air Force. Once, while testing a Mach 2 fighter-bomber at high altitude, the airplane rolled crazily and then whipped into a spin, upside down. Slayton regained control, but fought the plane in a careening fall that lasted nearly 30,000 feet. Alan Shepard has suffered engine flameouts at very high altitudes. Gus Grissom and Wally Schirra are Korean air combat veterans, as is John Glenn, considered a brilliant test pilot, with more than 5000 hours of flying time. Neil Armstrong remembers being nestled under a B-52 mother ship at 40,000 feet in the gleaming black rocket bird. He knows the changing purple darkness of the thin air at the door to

1

ILLUSTRATION BY KEN HODGES

3

4

outer space. He knows his reactions to the commands of countdown. He knows the command: "Drop minus one minute!" The instruments before him are obeyed as easily as one breathes. Perhaps someday, his skill with instruments and flying machines will be a thing of destiny, to a whole world watching him fly a Lunar Module . . . like the X-15 . . . "Check the pressurization cooling, the hydraulic pressure; reset generators, now the prime switch on. Arm the ventral-fin jettison. Everything OK . . . five . . . four . . . three . . . two . . . one . . . drop!" But the sounds of the X-15 are giving away to the sounds of Apollo, with the whole world a witness at it's birth hearing the words: "We have an ignition!" Now, stop a minute. You have completed your training, and you are an astronaut. You have learned how to put on your intravehicular space suit (**1**) in zero gravity. Like Buzz Aldrin, you learn to work the docking collar (**3**) on an Agena Target Docking Vehicle, while under water, in a zero-gravity simulation; or perhaps with a mountain guide and buddies (from left) Deke Slayton, Dave Scott and Scott Carpenter (**2**). You know what the Congressional Space Medal of Honor (**4**) is; what it's like to live in a cramped capsule from the aquanauts (**5**); even of the post-flight trailer (**6**). And, as an astronaut—on your way to the heavens at last—you know that there may be serious, even fatal consequences. But you believe deeply that eventual victory of man in space will come from mankind's resolve; by those who carry that torch—with their training.

2

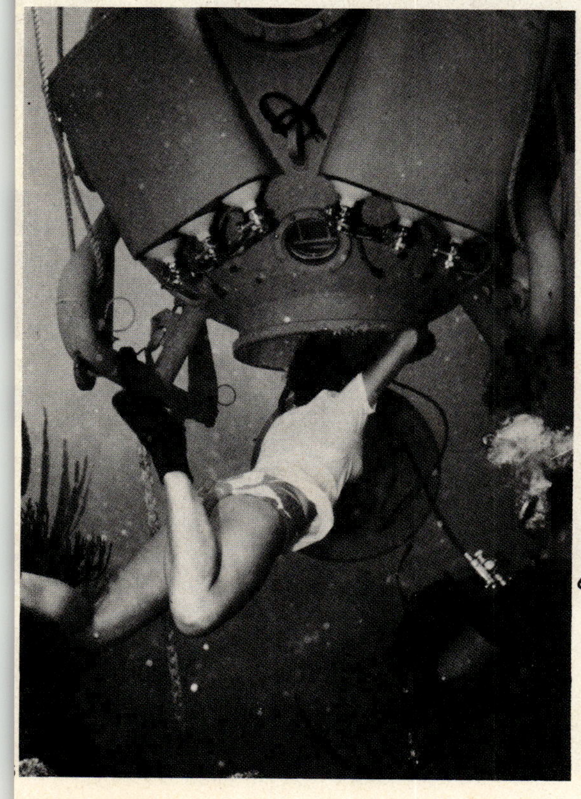

6

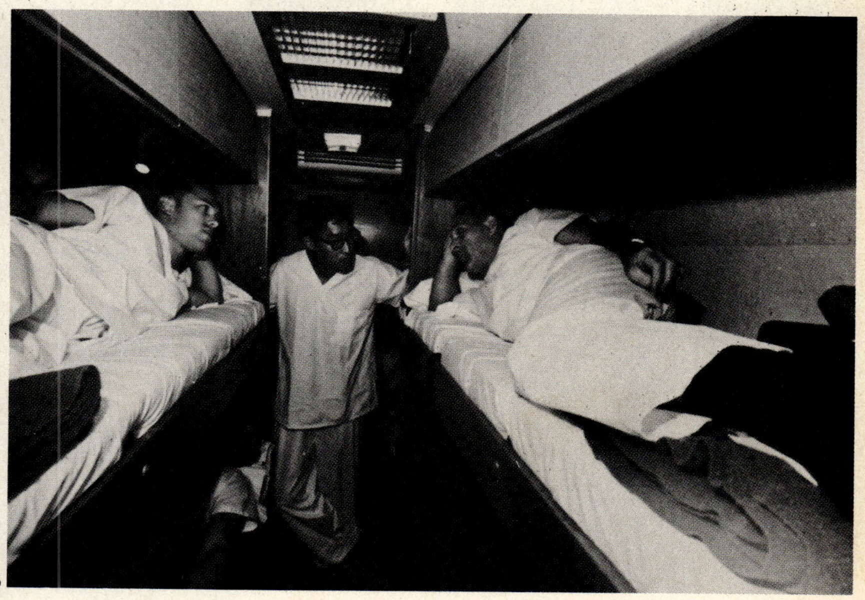

"A body in motion tends to remain in motion..."—so states Newton's second law of motion. His calculations and theories today solve some of the most difficult space travel problems but NASA is still left with the hardest one: How does one initially get the "body" (read "payload") into motion? Basically, the factor that must be overcome is gravity; the solution, reduced to its simplest term, is "thrust". If man is to someday walk among the stars, enough thrust must be generated to lift him beyond the grasp of gravity and place him into orbit. Some portions of the enigma yield readily to man's ingenuity, and by the early '60's there is already some progress toward the ultimate solution, but the goal of placing hundreds of tons of payload into orbit is still a dream. The old rigid-frame rocket is abandoned in favor of the Atlas-type missile construction. In this new design the cylinder wall is constructed of high-tensile stainless steel and its thickness is reduced to 19/1000 of an inch. It is held rigid by fuel pressure from wi-

UPRATING THE SATURN

The Saturn gets more push, a new look and the title 1B (for One Better?)

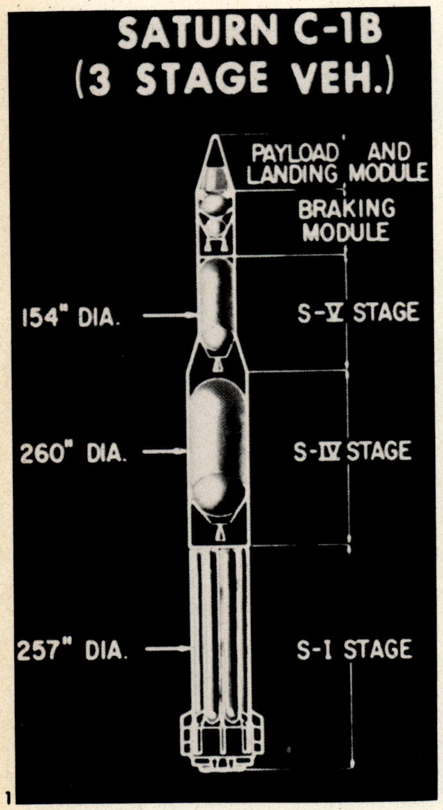

1

2

3

thin—a concept borrowed from a child's balloon. Though successful, it launches a payload of only a few tons and it is recognized that a larger, more powerful vehicle is necessary. The Saturn 1 rocket is now envisioned as the latest building block in the chain of interim answers. Three developmental steps are required before the Saturn-1B is evolved under the direction of Dr. Wernher von Braun at the George C. Marshall Space Flight Center at Huntsville, Alabama (**1**). The first stage, utilizing the new and more powerful H-1 Rocketdyne engines, consists of eight tanks clustered around an inner core tank. The rear propulsion bay measures 22 feet in diameter and is capable of producing 1.5 million pounds of thrust. The first Saturn-1B is to fly on February 26, 1966 and an evaluation of the launch vehicle is given by Dr. George E. Mueller, NASA Associate Administrator, who states, "The uprated Saturn . . . is capable of placing 40,000 pounds into orbit . . . " Launching the powerful vehicle presents some peculiar problems and launch Complex 34 is not entirely suitable for the Saturn-1B, so launch Complex 37 is designed to accommodate the behemoth (**2**)—the actual launch pad is similiar (**3**). A detailed breakdown of the first stage of the Saturn-1B (**4**) shows design details and tank arrangement. Actual construction/assembly (**5**) is carried out at the Chrysler Corporation assembly area. The first of the Saturn-1B vehicles is lifted into place at the Cape's Complex 37 (**6**) in preparation for the first historic launch in 1966.

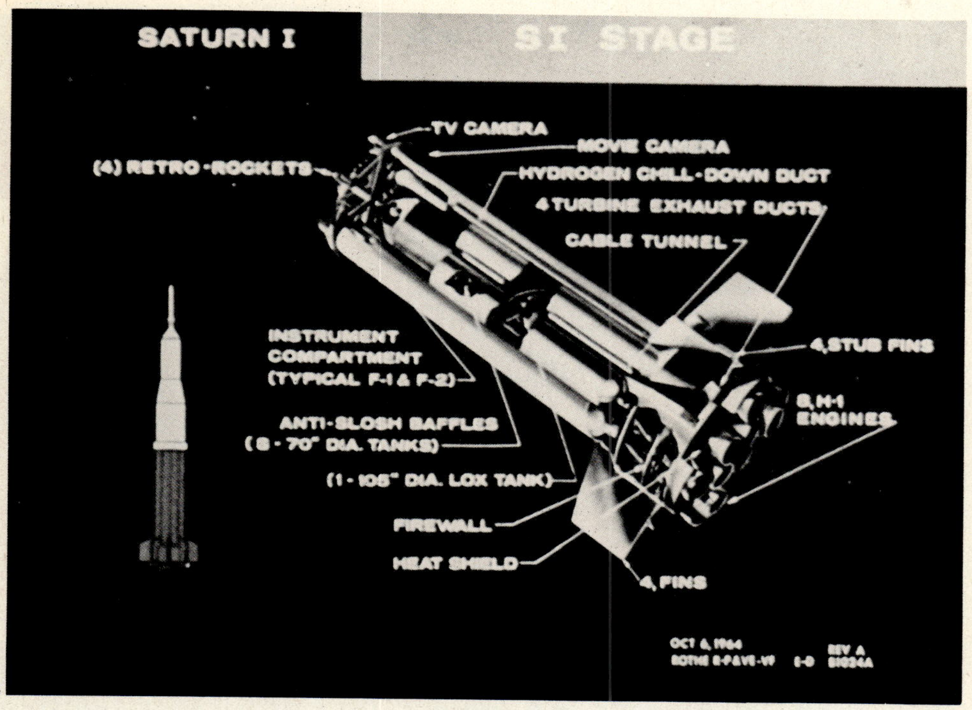

4

5

6

UPRATING THE SATURN

The first flight of the Saturn 1B will not only be a test of a new and more powerful first stage engine, but it will also be utilizing a new second stage: the S-IVB (**1**). This has a completely new engine design that bears the nickname "Little Joe" but which is more properly designated as the J-2. Earlier Saturn 1 test launches used a cluster of six liquid propellant engines that had a combined thrust of only 90,000 pounds—the new engine, produced by the Rocketdyne Division of North American Aviation, uses a fuel of liquid hydrogen and a liquid oxygen oxidizer to produce a thrust of 200,000 pounds. It is the largest liquid hydrogen/liquid oxygen engine in the world. An overall view of S-IVBs (**2**) in the Vertical Assembly Building at Cape Kennedy gives some idea of the size of the vehicle in relation to man. This building, incidentally, is one of the tallest and biggest structures in the world measuring 716 feet long, 518 feet wide, and 52 stories (526 feet) high—only a few feet shorter than the 555-foot Washington Monu-

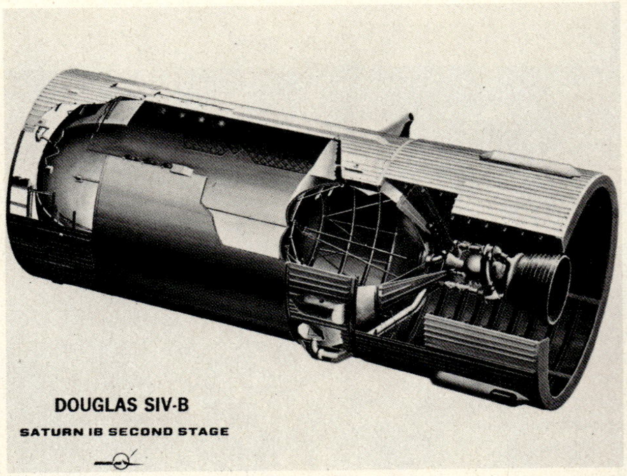

1

PHOTO COURTESY MCDONNELL DOUGLAS

2

3

PHOTO COURTESY MCDONNELL DOUGLAS

4

5

ment. The aft and forward adapter skirts that link the S-IVB with the aft lower stage and with the forward Instrumentation Unit of the space vehicle are also seen (**3**). In the background is a completed S-IVB with the aft skirt to the left and the forward skirt to the right in the foreground. Upon construction completion at the Douglas Aircraft Company in Huntington Beach, California, the S-IVB is loaded into an Aero Spacelines Super Guppy (**4**) for transportation to Cape Kennedy. The second stage weighs approximately 58,500 pounds with instrumentation. The first test flight goes off as programmed and the S-IVB, during a seven minute burn, boosts the payload Service and Command Module into a low earth orbit. During the second test flight, which takes place on July 5, 1966, the S-IVB is placed into orbit as planned and successfully tested for orbital restart capabilities. This is tremendously important for future planning, as the S-IVB will be used as the third stage of the Saturn V rocket with which men hope to reach the moon. After several orbits on the second test flight, the S-IVB is subjected to exceptionally high gravity loadings during pitch, roll and yaw maneuvers. It exceeds expectations and only breaks down after far surpassing design predictions. Preparations for the third test flight are made (**5**) with the pre-mating of the S-IVB and its adapter skirt at the Cape. The first and second stage of the rocket are stacked (**6**) and the test flight is launched. The S-IVB performs perfectly, placing the payload into orbit precisely.

PHOTO COURTESY MCDONNELL DOUGLAS

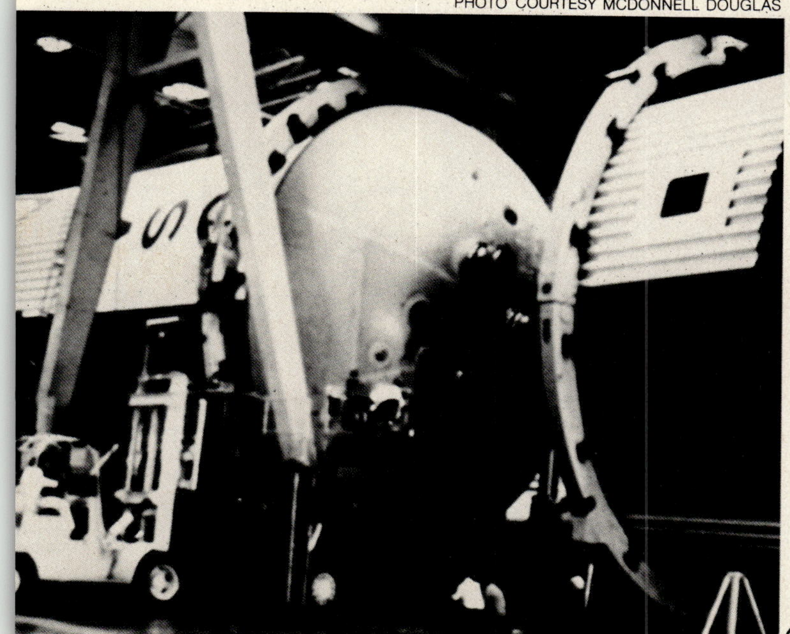

6

UPRATING THE SATURN

Equally important as the propulsion system to the success of the Saturn 1B is the Instrument Unit—the vehicle's brain. Designed by IBM, the unit utilizes the most advanced state-of-the-art electronics, and the complex computer portion programs the entire sequence of flight events from high atop the second stage of the Saturn 1B. In the shape of a ring, the Instrument Unit can be examined by considering it to be divided into four sections, each one of which contains thousands of feet of wire and many thousands of micro-miniature electronic components (**1**). Control of the spacecraft becomes almost monumental in complexity because in the zero-gravity environment every inertial load change or shift must be sensed by accelerometers, analyzed and reacted to by the brain. Even the possible sloshing of the liquid propulsion fluid must be dampened and baffled to prevent a deviation to the pitch, roll or yaw axis and a subsequent loss of trajectory stabilization. The solution, and the success of the American space ef-

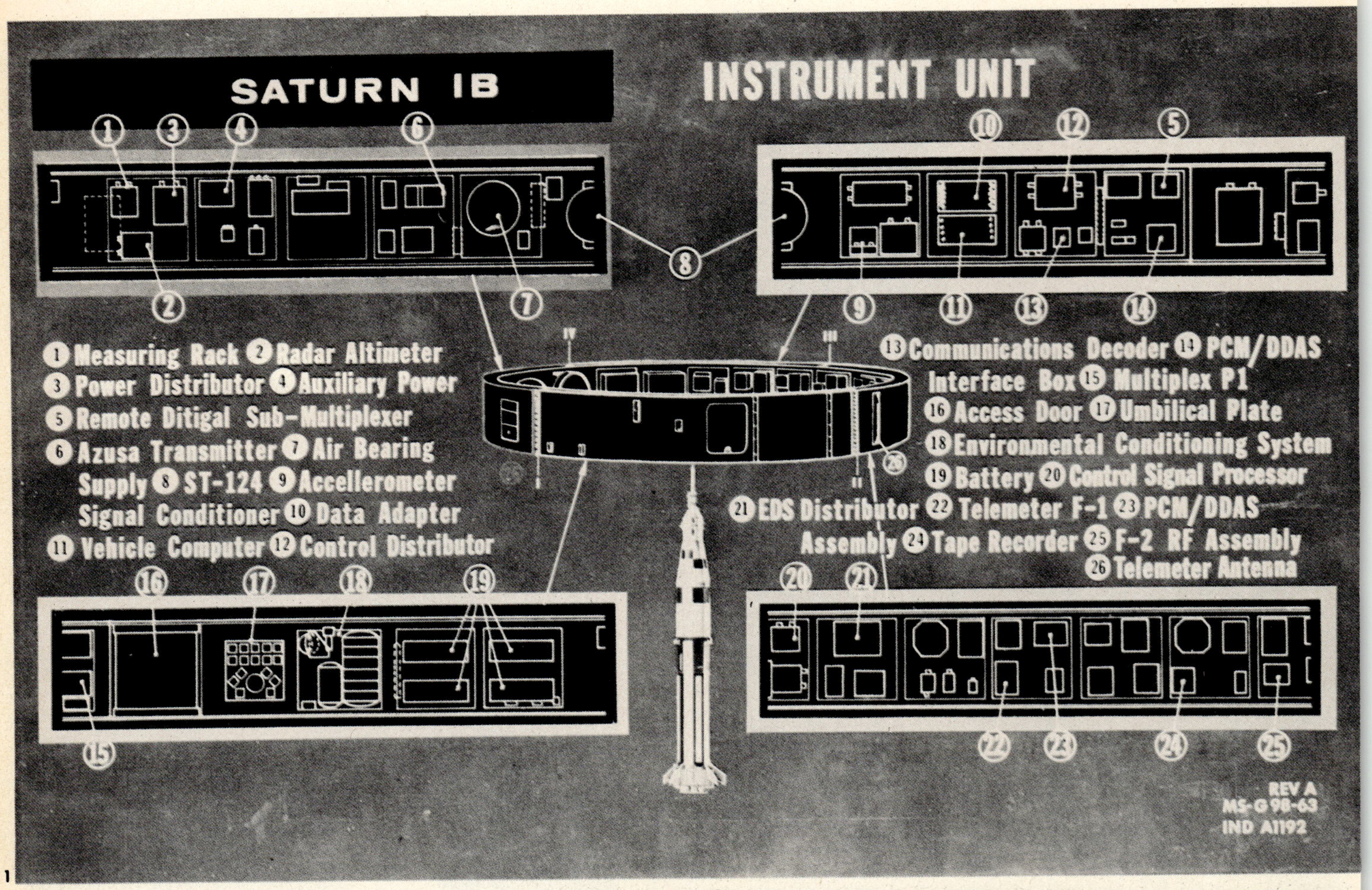

1

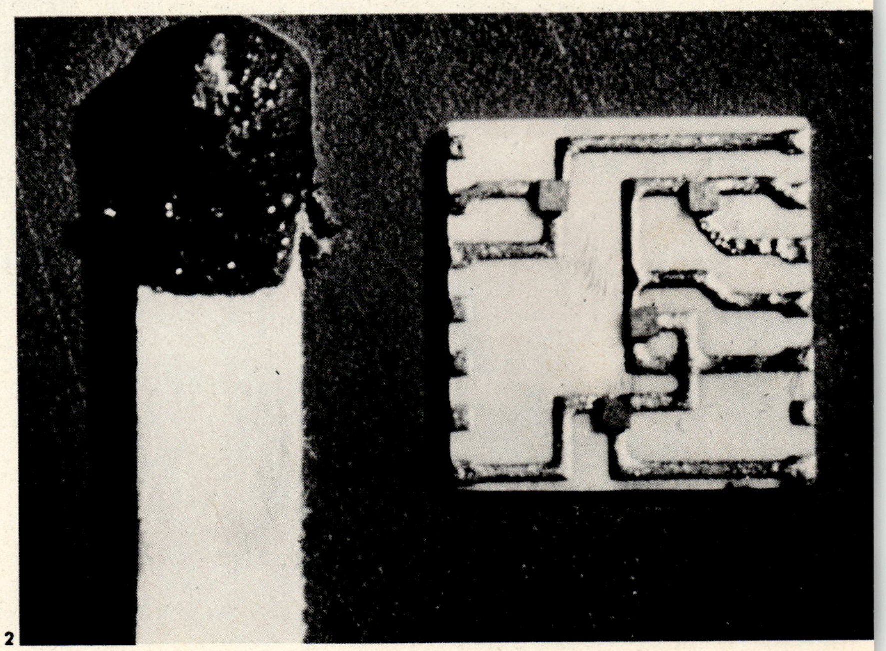

2

fort, lies in the field of micro-miniature electronics as much as in the field of aerodynamics. Modern miniaturization techniques developed primarily under the democratic theory of industrial competition far out-distance the efforts of our nearest competitor, the Soviets. But U.S. officials believe that the Russians are developing a rocket with equivalent power to the forthcoming Saturn V—it is also believed, however, that they lack the high-energy upper stage rockets such as the S-IVB and the control instrumentation required to maneuver such a vehicle. In the normal course of daily business events American electronic manufacturers routinely produce modern miracles of micro-miniaturization. The equivalent of hundreds of electronic circuits can be contained on a small chip not much larger than a match-head (**2**), and plug-in cards, measuring 2½x4½-inches and containing as much circuitry as 64 television sets, are being manufactured daily on an assembly line basis. An entire circuit board—a part of the coreplane memory of the Saturn computer and containing thousands upon thousands of circuits—is manufactured under "clean-room" conditions at the IBM Huntsville installation (**3**). A complete interconnection board is shown (**4**) together with the miniaturized memory drum of which it is a part. The completed Instrument Unit is now transported to the Cape, in a Super Guppy (**5**), and after testing at Kennedy the Unit is hoisted up and placed on the top of the S-IVB stage of the stack (**6**). It will be jettisoned with the S-IVB.

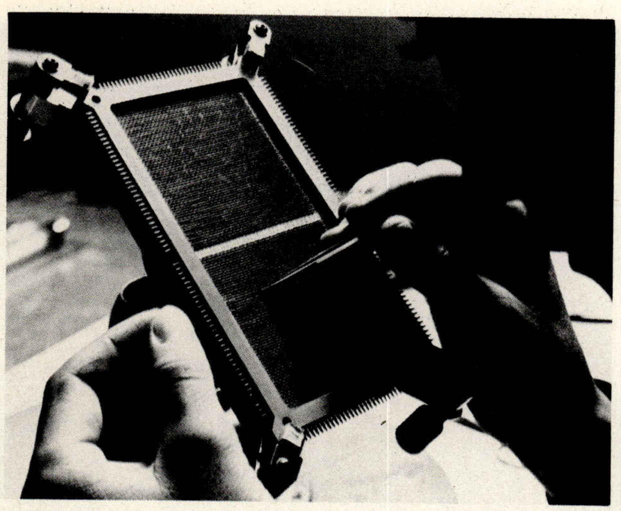

3

4

5

6

UPRATING THE SATURN

The first complete S-1B rocket is christened the Saturn/Apollo-201 (SA-201) with the planned future shots designated as the SA-202, 203, 204, etc. Preparations get underway weeks in advance and the Apollo spacecraft, the payload for this launch, undergoes an alignment test (**1**) to ensure flight readiness. This unit is next one up in place on the rocket stack after the Instrument Unit and the lunar adapter are attached. As the tower is rolled away (**2**) the entire vehicle can be seen: the lower first stage, the S-1VB second stage with the IU above it and the lunar adapter next, then the Command/Service Module. The stack is capped by the launch escape tower. On February 10, SA-201 completes a countdown demonstration on Pad 34 in preparation for a scheduled February 22 suborbital flight test (**3**). However, this launch is postponed due to cloudy weather conditions that would hamper visual and photographic coverage. Looking up from below (**4**)—note the vertical shutters on the right. These are in-

1

2

stalled on Complex 34 and 37 to afford protection from possible hurricane damage. The entire vehicle is prepared on its launch platform and on February 26, 1966, the Saturn/Apollo 201 blasts off! (**5**) The mission is almost aborted, however, because the S-1B stage indicates a low hydrogen pressure condition. The launch controller notices the malfunction and recycles the countdown—an intermittant fault apparently—this time the lift-off and powered flight go as programmed. After ignition of the second stage, the multi-engined first stage is jettisoned. After a seven minute burn the second stage and the IU separate from the Apollo spacecraft (no. 009) placing it into a orbit at an altitude of 310 miles. The purpose of this test is to qualify the heat shield of the CSM combination and certain major spacecraft subsystems. The test launch is successful and the descent is equally successful, with the CSM firing reaction-control system rockets for 30 seconds to increase the spacecraft's re-entry speed. The Service Module is jettisoned and the Command Module re-enters at 27,500 fps—a speed higher than desired but necessary to test the heat shield's ability to withstand a re-entry temperature of approximately 4,000 degrees F. The spacecraft splashes down in the Atlantic at 11:15 AM EST within 35 miles of target. Astronauts Gordon Cooper and Don Eisele watched the SA-201 launch (**6**), and with considerable satisfaction later read the NASA announcement: "All primary mission objectives of first Apollo/Saturn 1B...were achieved."

5

6

UPRATING THE SATURN

After the success of the SA-201, preparations are made for the launching of the next test vehicle. Due to a change in the sequence of information requirements this is SA-203 instead of SA-202. The shot is basically an engineering study of the liquid hydrogen fuel behavior, a test of the launching of the S-IVB as a second stage entity and a test of the Instrument Unit. SA-202 is completed at Complex 34 (**1**) while the erection of SA-203 is carried out at Complex 37 (**3**). The first stage of SA-203 is moved into place (**4**) and the S-IVB is carefully lowered into position (**5**). SA-203 is launched on July 5th from the ETR (**6**) and the two-stage Saturn 1B throws the unmanned payload, consisting of the S-IVB second stage, the IU and a nose cone (**2**) (there is no Apollo module on this flight) into circular orbit at 117 miles altitude. The S-IVB engine burns briefly in the atmosphere, then it coasts. The engine's ability to restart is demonstrated, but no reignition is attempted. The liquid hydrogen fuel behavior is photographed and

1

2

3

the pictures transmitted to four ground stations by TV. During the fourth orbit differential pressure tests are performed and after building up internal pressure in the second stage—far in excess of design specifications—the S-IVB, shatters but all mission objectives have been achieved. The 58,500 pound, 93-foot, S-IVB/IU-nose cone combination is the heaviest satellite ever orbited by the U.S. Preparations proceed at Complex 34 and SA-202 is launched on August 25th at 1:16 p.m. EDT. The Saturn-1B launch vehicle hurls the 56,000 pound unmanned Apollo spacecraft (011) into suborbital flight. This is the third flight test of the Saturn-1B, the second performance check of the Command Module heat shield and the second flight test of major spacecraft systems. The Service Module propulsion engine burns for 3 min. and 35 sec. after spacecraft separation and blasts the Command Module to an altitude of 706 miles. The Service Module engines are ignited three more times to test the ability for rapid restarts, and on the last ignition the Service Module separates from the Command Module. Re-entry speed of the Command Module is greater than 19,900 mph and the spacecraft's exterior reaches a glowing temperature of approximately 2,700 degrees F., but temperature on the interior of the capsule remains at 70 degrees F. The spacecraft splashes down in the Pacific Ocean (**2**) about 500 miles from Wake Island—an error of some 200 miles—at 2:49 p.m. EDT., and recovery is made by the USS *Hornet*.

6

APOLLO 1

Though not the first spacecraft in the Apollo series, this will be the first to be flown with men in it. Three human beings will fly in a complex machine designed to perform all manner of things still considered in the realm of mythical acts by half the world. Though as Gus Grissom can well testify, things do go wrong. A wealth of spacecraft knowledge has been brought to fruition for this mission, and the suits upgraded from the Gemini program; no one, however, has done what these men intend. No one has flight-tested this new capsule in which dozens of interlocking subsystems must achieve perfect synchronization in the solar wind, unprotected by the thin film that is earth's gaseous skin. Still these men are not deterred by past problems so much as they are inspired by present and future possibilities. Symbols of their nation's corporate and technological might, they carry America's banner with them toward the oceans of the moon. Their own insignia says as much with utter simplicity: Apollo flies above the place which gave it birth, nose aligned with its final destination. Grissom, White and Chaffee: they've spent some of the most important years of their lives training for this, their special moment on a Florida seashore.

VIRGIL I. GRISSOM

COMMAND PILOT/ Born in Mitchell, Indiana, on April 3, 1926, Grissom received a BS degree in Mechanical Engineering from Purdue University. A USAF Lieutenant Colonel, he has logged 4600 hours flying time, of which 3500 have been in jet aircraft. One of the seven original Mercury astronauts, he piloted the second and final suborbital Mercury test flight in 1961. In 1965 he served as command pilot on the first manned Gemini flight; subsequent to this assignment he was the back-up command pilot for Gemini 6.

EDWARD H. WHITE II

SENIOR PILOT/ Born on November 14, 1930, in San Antonio, Texas, USAF Lt. Col. Ed White received a BS degree from the Military Academy at West Point in 1952, and a MS degree in Aeronautical Engineering from the University of Michigan in 1959. He spent three and a half years flying F-86s and F-100s in Germany, and has more than 4200 hours flying time—3000 of them in jet aircraft. As pilot for Gemini 4 in 1965, he performed the first extravehicular activity in the American manned spaceflight program.

ROGER B. CHAFFEE

PILOT/ A native of Grand Rapids, Michigan, where he was born on February 15, 1935, Roger received his BS degree in Aeronautical Engineering from Purdue University in 1957. After graduation he entered the Navy, and was assigned to Heavy Photographic Squadron 62 at the Naval Air Station in Jacksonville, Florida. He now holds the rank of Lieutenant Commander, USN; of his more than 2300 hours flying time, in excess of 2000 have been in jet aircraft. He has two children, and is a gun collector. This is his first space flight.

GRISSOM
WHITE
CHAFFEE
APOLLO 1

APOLLO 1
Preparation

In late 1966, NASA announces the flight crew for the first manned Apollo. Astronauts Grissom, White, and Chaffee (left to right) beam at their selection for this premiere opportunity (**1**). They will travel through a new ocean, an adventure which in their own childhoods was still a dream, in a craft which is the most complex piece of work yet devised by man. To operate in it requires that they shed the dark suits and ties of the ordinary world in favor of new skins—each one appearing almost as precisely tailored to its wearer as the one that he was born with. The project requires a new hat, too, as Grissom demonstrates (**2**). The hose on his left side pumps in oxygen, the pressure of which is modulated by the hose on his right; the electrical and communications system enters through the curling cord in his chest, more clearly seen on Ed White (**3**). The chestpack contains an inflatable vest for use in emergency water egress should that be necessary at splashdown—as it was when Grissom's Mercury spacecraft, *Liberty Bell*

1

2

7, sank beneath him. Their suits, an updated version of those worn in the Gemini program, must be put on and taken off repeatedly as the astronauts become accustomed to working with them. Grissom (in foreground) and White are seen adjusting their gloves as carefully as though preparing for an evening in white tie and tails (**5**). Most certainly the occasion demands an outfit as rigorously tailored and precisely fitted as any tuxedo; but its rules of conduct are far more stringent than those of earth-bound evenings, and the penalty for ignoring the etiquette of space infinitely more severe. The astronauts are trained and the craft designed so that any one of them alone can maneuver it back from space. But before they even reach that phase of the mission, they must learn to use the spacecraft with as much familiarity as they would a lawnmower on earth. With each succeeding stage of the manned spaceflight program, NASA and the contractors make modifications in design of the vehicle, and to the multitude of subsystems which add to the safety of its crew. To learn of these new modifications, the crew spends weeks in the spaceflight simulator (**4**) seated as they will be in the capsule—White in the middle with Chaffee on his right and Grissom to his left. Not all the procedures are interesting and some of them—like pressure testing in the centrifuge—are painful. The back-up crew, astronauts Eisele, Cunningham, and Schirra (**6**), likewise trains against the possibility that the main crew will be unable to fly.

3

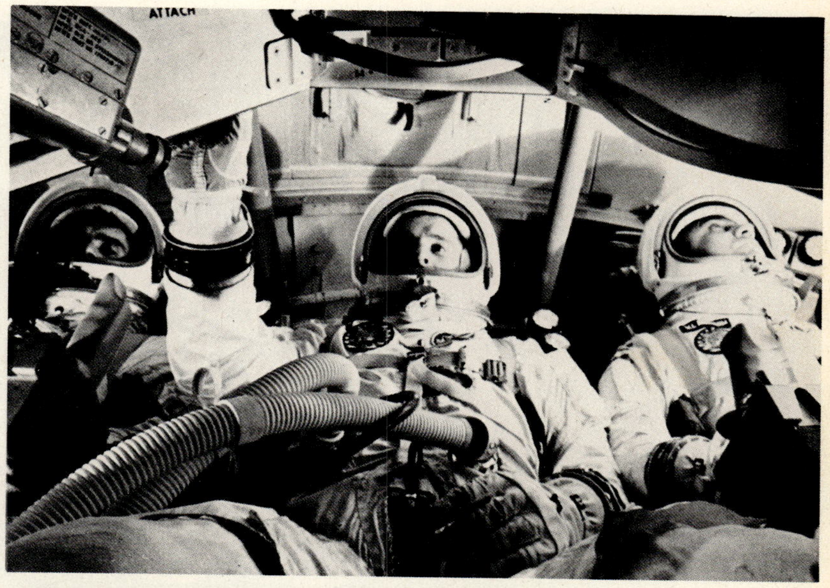

4

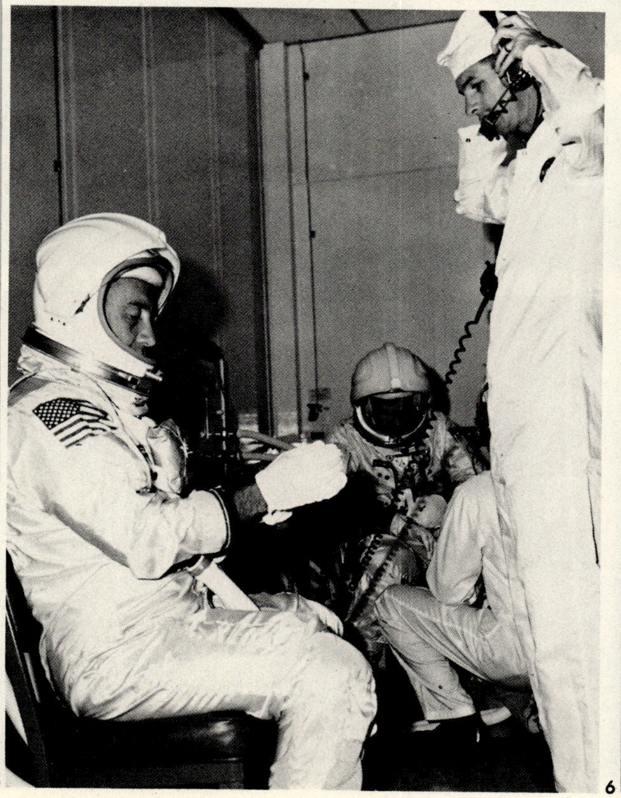

5

6

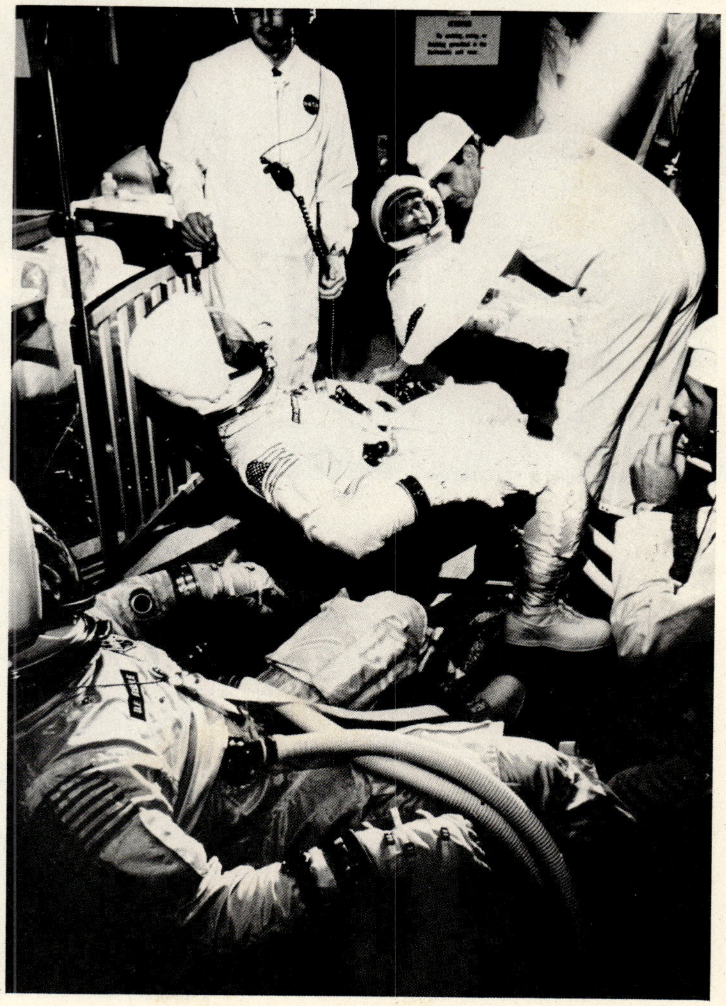

APOLLO 1
Pre-launch

Absolutely every process of the mission's mechanics must be practiced and repracticed—the same sort of repetitive drill necessary in learning to swim or sail a boat. Easily as important as learning to fly the Apollo spacecraft is training to get out of it. For this the crew goes to a boilerplate capsule abob in a swimming pool. Though the boilerplate is of the same size and weight as a fully functional Apollo module, it lacks the complex interior gadgetry of the simulator. Assuming the capsule has just splashed down (**1**), command pilot Grissom would be looking out the window to the left of the hatch door, pilot Chaffee would be at the window just visible to the right, and senior pilot White would be positioned between them at the hatch door. Since all are lying on their backs, the hatch is behind and above White's head. Thus he has naturally been the first to exit; were he unable to do so, the others must crawl over him—in a space more cramped than a compact car. While White floats in the foreground, Chaffee

1

2

gingerly steps onto another life raft, remembering that he might as easily be in 1000 fathoms of ocean water as the 10 feet of the pool—and could drown in either. After innumerable practices at water egress, the crew moves to still more practice with an actual capsule. Here (**2**) we see Grissom, Chaffee, and White posing beside a Block 1 vehicle. Unlike capsules designed for later missions, those of the Block 1 type have no docking tunnel; it isn't needed since this is to be a simple earth orbital mission to check spacecraft systems. In this group shot Chaffee, at least, is still capable of a broad grin despite the grueling procedures. One must be just as careful and surefooted getting into the capsule as getting out; to that end back-up commander Walter Schirra is seen during ingress practice, approaching the hatch door of a taped and covered capsule (**3**). Gus Grissom goes through the motions with calm deliberation (**4**), his face aglow against the darkness which surrounds him. It is he who will normally operate the craft's flight controls. The astronaut in the center couch has guidance and navigation as his principal tasks, though he will also take over flight duties at times; were this a lunar mission it is he who would remain in the CM while his comrades descend to the moon. Occupying the right hand couch, pilot Chaffee is responsible for management of spacecraft subsystems. Learning all this, and continually running through it, is taxing of course, so the crew takes a well-deserved break (**5**) for a moment of contemplation.

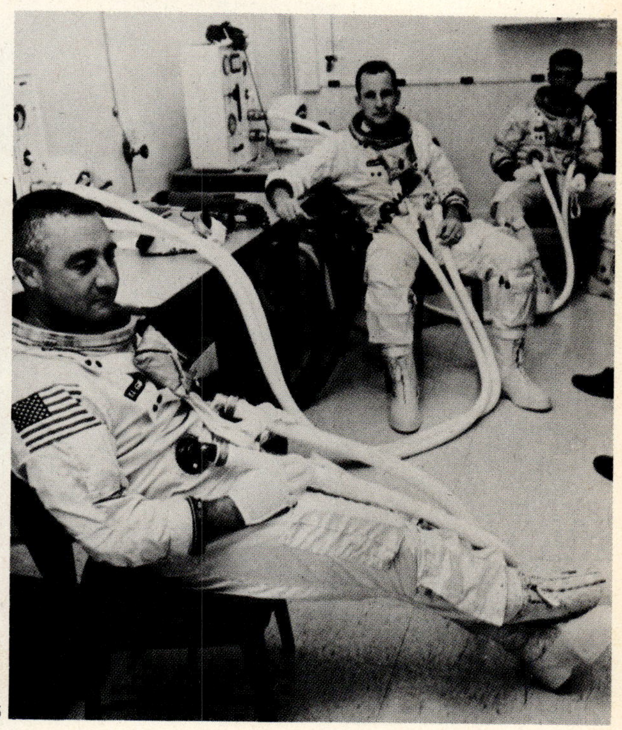

5

APOLLO 1
Pre-launch

After the crew has been sufficiently familiarized with their spacecraft through practice with the simulator, the time comes for practice with the combined capsule and launch vehicle. Those subsystems for which Chaffee is responsible are the major ones comprising communications, earth landing, environmental control, electrical power, guidance and navigation, launch escape, reaction control, thermal protection, and stabilization. They must now be connected and thoroughly checked. In space the margin for error is nil. Work on capsule No. 012 begins in August 1964; the basic structure is completed the following autumn, and subsystems are assembled and installed by March 1966. With its final reviews completed by October 1966, there is a possibility of launch in December or January; but failure of the oxygen system regulator in mid-October and problems in the Environmental Control Unit force postponement. Eventually Apollo spacecraft No. 012 is hoisted aloft (**1**) festooned with tags and ribbons that mark parts to be

1

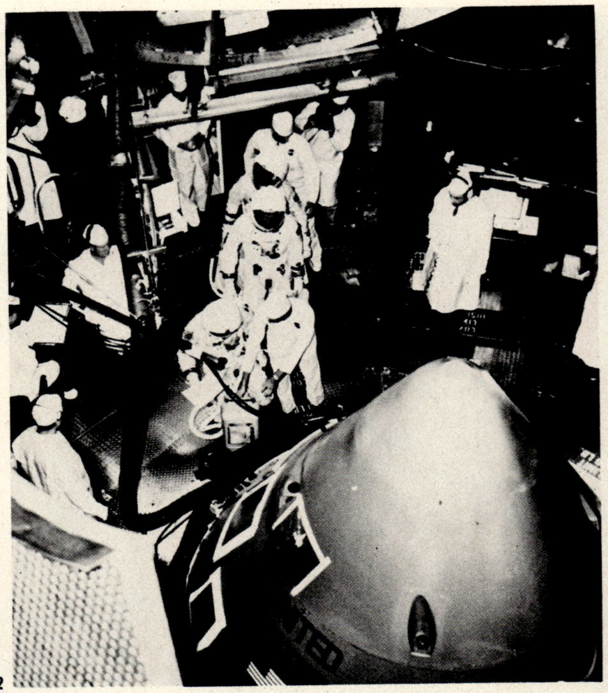

2

3

4

checked before flight readiness is achieved. Meanwhile the Saturn S-IVB second stage (**3**) is lifted atop the Saturn S-1B first stage (**4**) at Launch Complex 34. There is no third stage because it is not going to the moon. This combined Apollo-Saturn, No. 204, stands 224 feet tall; the mobile service tower enclosing it is 310 feet high, weighs 3500 tons, and has seven levels and eight movable platforms. The mating of the CM with its 24-foot Service Module (SM) is among the most delicate and taxing of pre-launch preparations. Seen here (**5**) suspended in the Manned Spacecraft Operations Building above the 13,475-pound SM, spacecraft No. 012 shows its electrical viscera: 20 miles of cables, wires, and tubing . . . and potential trouble. The protrusion on the right of the picture is part of the reaction control subsystem; four such devices will guide the spacecraft in flight. The capsule is covered with a layer of ablative material designed to dissipate the heat of re-entry by burning away. This layer covers a two-shelled hull—the outer shell of which is formed of stainless steel honeycomb braced between stainless steel sheets. The outer shell is separated from the aluminum honeycomb inner one by an insulation space filled with two layers of microquartz fiber. This metallic skin is the only thing protecting the interior from the full blast of raw solar energy and speeding micrometeoroids, at an altitude where a man's blood would boil and his skin burst like a stewed tomato's were he exposed to the fury of airless space and the naked sun.

5

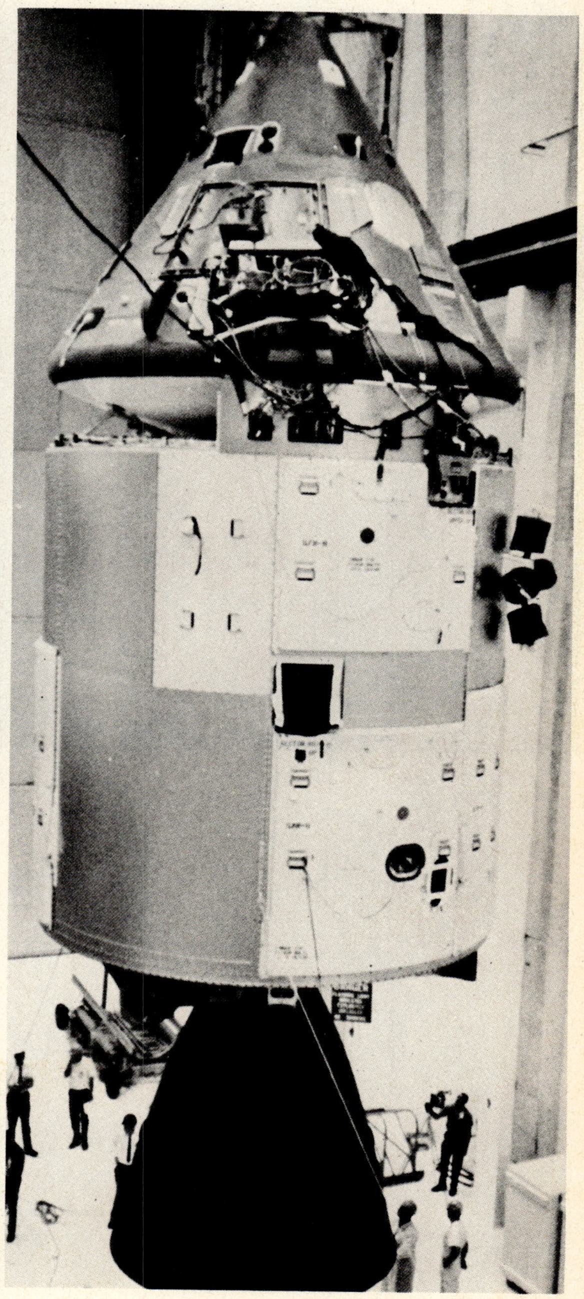

APOLLO 1
Post-flight

On January 27, 1967, at 7:55 AM EST technicians at Kennedy Space Center begin a test of spacecraft No. 012, in which the vehicle's systems are to be put through a simulated countdown and launch. The crew, fully suited up as though for an actual flight (**1**), strides through winter sunlight to the great tower enclosing their craft. At 1 PM they enter, Grissom going first, followed by White and finally by Chaffee. These are the positions they must hold through all the long hours of the afternoon—much of it the dull, listless times between checks of various systems. In one such lull before the hatch is closed for the final time, we see White calmly waiting, almost as though asleep, in the clear light filtering down through the open tower (**2**). Grissom appears steeled for the next development, and Chaffee in the background stares into space, for this is but one of many routine tests. Countdown proceeds until Grissom notices an odor in his space suit oxygen supply; the count is resumed when no impurities are found after sampling.

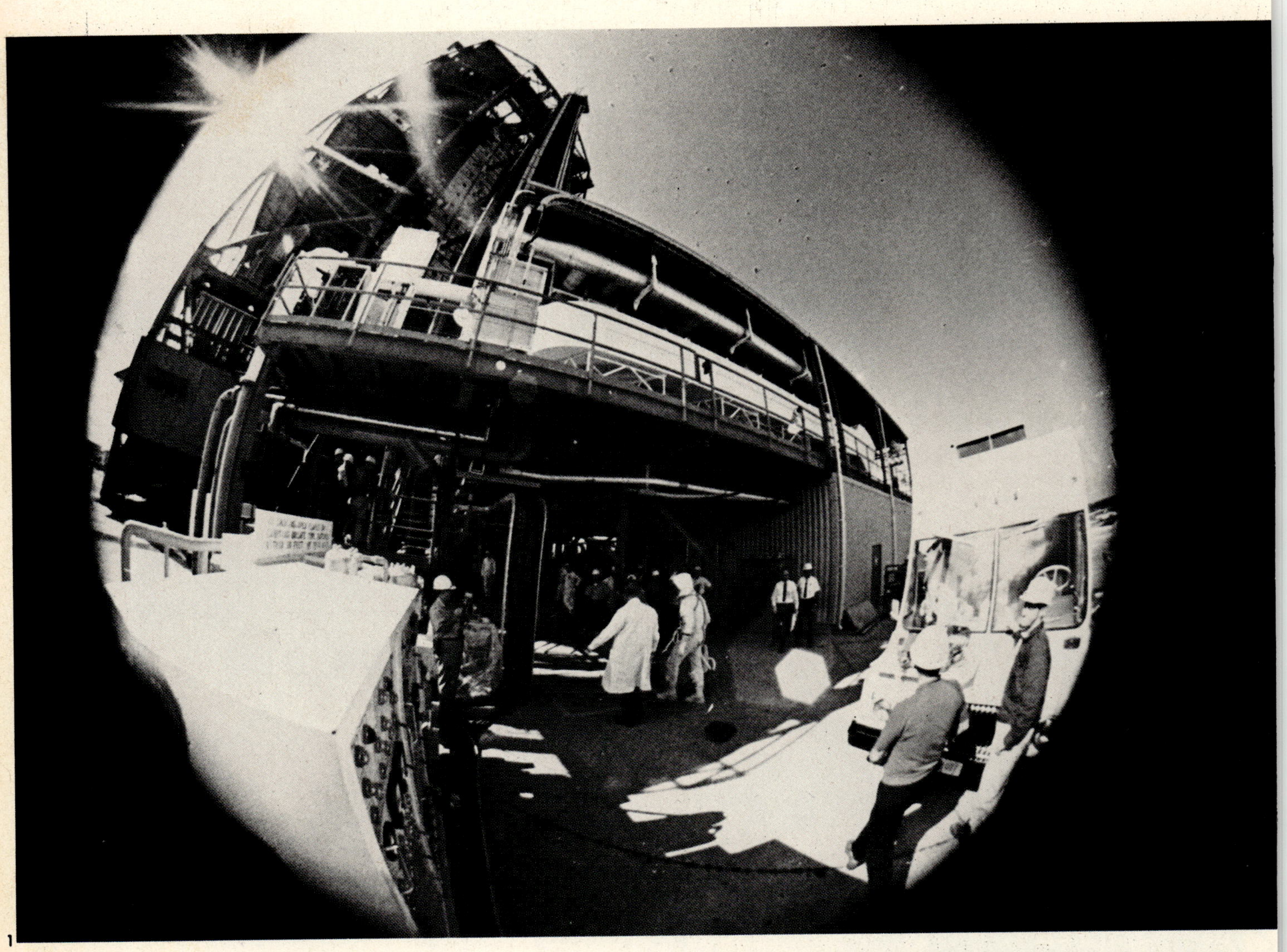

1

4

The cabin itself has been filled with oxygen at a pressure of 16.7 psi. In an actual flight, pressure in the cabin would be dropped to 5 pounds psi after launch, which against zero pressure outside will seal the hatch tightly. Shortly after 5 PM, there is trouble with communications between spacecraft and blockhouse; and the countdown must be held again at 5:40 PM. Though it is resumed, new problems appear in the entire communications network. Then at 6:30 PM there is a surge in voltage, followed at four seconds after 6:31 PM by a cry from within the sealed craft: "fire". Apparently the fire starts in a bundle of wires beneath Grissom's couch; burning with the speed of an explosion in the 100% oxygen atmosphere, a fiery wall prevents him from reaching a valve to vent it outside. A few seconds more pass and there is a cry of pain, as flames roar from the base of the craft. Pressure in the cabin has soared to 29 pounds psi; and heat to 2500 degrees. In 15 seconds the aluminum inner hull splits. The crew of Apollo 1 is mortally wounded, as the fire severs their suit-breathing loops and they suffocate. It takes some five minutes to break into the super heated capsule. But by then the men are dead. The scorched spacecraft is cordoned off (**3**) and the tragic examination begun. Amidst the gutted and twisted remains of the interior, the flight log curiously survives with little damage (**4**). The next day the capsule is removed, and lowered in its grey shroud (**5**). The United States of America was shocked!

3

At Chaffee's funeral (**1**), Lyndon Baines Johnson stands silently representing the nation. But this ceremony at Arlington Cemetery buries neither the country's grief nor its questions as to how such a terror could have blazed above the Cape. Within hours after the bodies are removed from spacecraft 012, NASA assembles the Apollo 204 Review Board to determine precisely the fire's cause and whom to blame. Neither objective is reached with absolute certainty. It is noted the plastic materials in No. 012 (**2**), while highly fire-resistant in normal atmosphere, burn readily in pure oxygen under pressure. The charge that this deficiency in design is the fault of the prime contractor, North American Aviation (NAA), must be balanced by the observation that NASA accepted it. In a room of the Pyrotechnic Installation Building at the Cape (**3**) members of the Board's supporting panels examine charred CM components; of particular interest is the severely damaged area near the floor in the lower forward section of the left

CAUSE & EFFECT

The A/S-204 fire investigation results in blame, shame and changes

1

3

hand equipment bay below the troublesome Environmental Control Unit (**4**). A view of this same area in spacecraft 014 (**5**), a Block I flight model very similar to No. 012, clarifies the area determined to be the most probable initiator of the fire—though destruction is too great to be conclusive on this point. A section of the DC power cable crossing over aluminum tubing is subject to bruising by the lithium hydroxide access door. In the picture of No. 012, a section of the power cable is in the center foreground, with the horizontal tubing coming in from the left; the door has been removed. Once again the charge of sloppy design on NAA's part must be counterbalanced by acceptance on NASA's. As astronaut Frank Borman, a Board member, notes, no one gave much thought to possibilities of fire in the spacecraft—least of all the astronauts themselves. Not only was the capsule a flammable deathtrap, but furthermore, tower personnel were unequipped to fight the fire. Revelations of this sort cause a furor in Congress, and hearings start immediately after release of the Board's report. But still it is not determined whether the blame is primarily NAA's, or NASA's, or some unknown worker's carelessness. Ultimately, the problem seems to lie in the size of Project Apollo (worth $2.2 billion to NAA by 1966), its sheer complexity and the brutal deadline of a lunar landing within the decade. At least the tragedy prompts changes in door design, materials, and launch team preparation. There is enough shame for everyone.

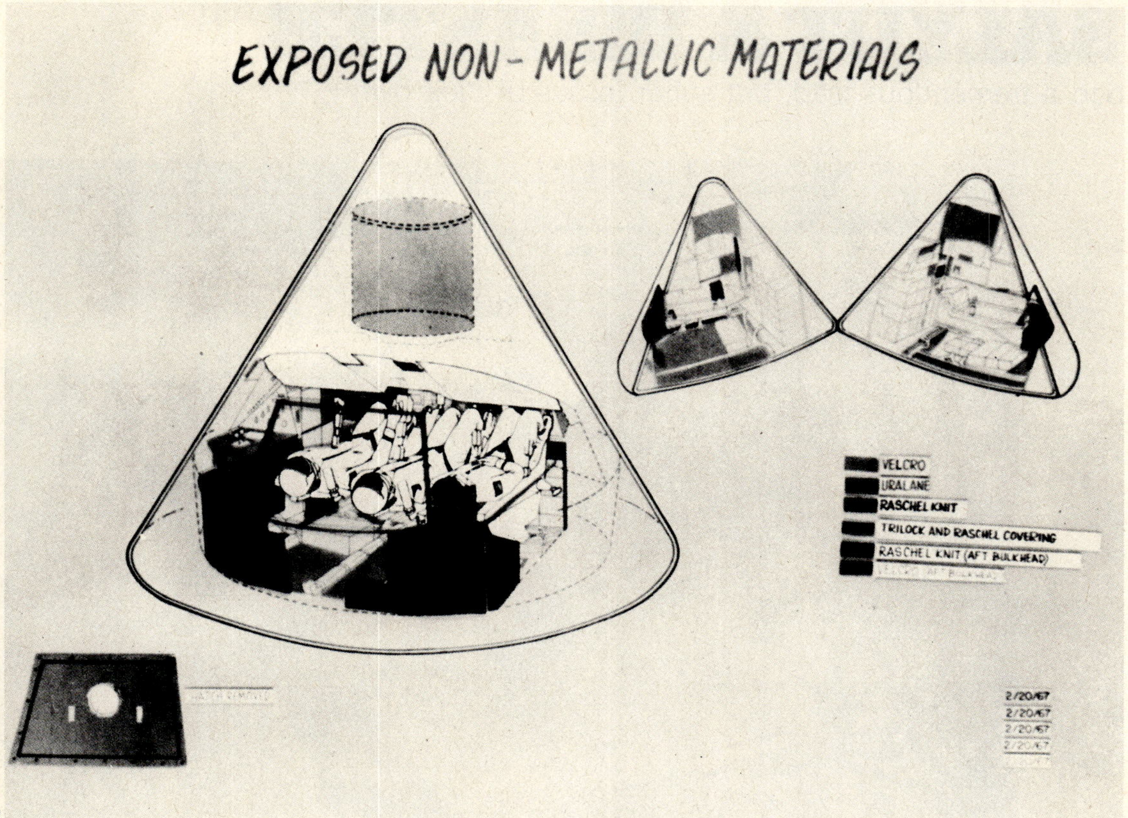

2

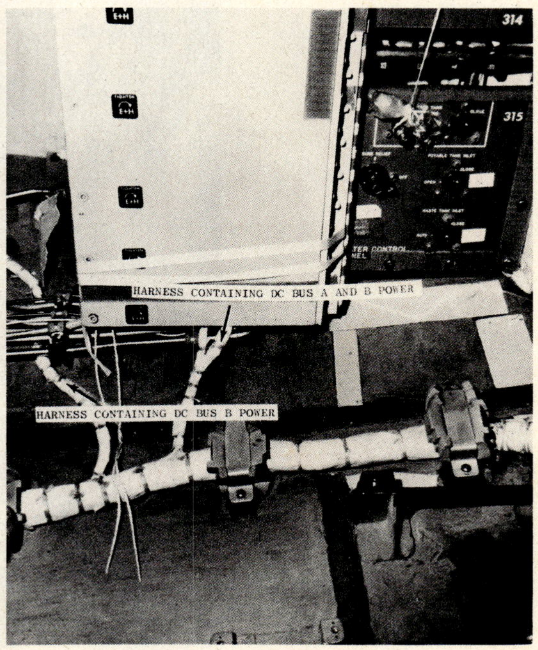

5

After the orbital flight of Voskhod 2 in March 1965, the Russian space effort appears to have taken a distinctly different path from that of the United States. While NASA has initiated its Surveyor and Lunar Orbiter programs, U.S. emphasis is directed towards Project Gemini and solving the problems of rendezvous and docking. Although the Russians do achieve the first soft-landing on the moon in 1966, as well as the first lunar orbit, their space program seems to be in low profile. But unknown to the West, they are preparing an eight-ton spaceship that can dock in earth orbit, either with a target vehicle similar to Agena or with another identical spacecraft. While technical difficulties during 1966 prevent them from orbital docking (thus the apparent emphasis on non-manned launches), these problems appear to be solved by early 1967 and the flight of Soyuz ("union") 1 is scheduled for April, about two months after NASA intends to launch Apollo 1. But the Apollo mission ends in tragedy a month before !aunch date

SOYUZ, SALYUT & DISASTER

The Russians are handed a tremendous lead, but suffer losses of their own

1
COURTESY TASS FROM SOVFOTO

when a flash fire in the Apollo capsule claims the lives of its crew during a routine test. While Soyuz 1 is being erected at the Baikonur Cosmodrome near Tyuratam, USSR (**1**), the Russian press chides NASA, charging haste as causing the accident. Cosmonaut Col. Vladimir Komarov (**2**) is selected to pilot the first flight of the Soyuz spacecraft, and on April 23, 1967 at 3:35 AM Moscow time, he is launched into orbit. Just 25 hours and 18 orbits later, Komarov plunges to his death when Soyuz 1's landing chute tangles during re-entry and fails to slow his spacecraft down. While the official Soviet news agency Tass maintains that all went well with Soyuz 1 until its final moments, veteran observers of the Russian space program suspect the following sequence of events: Komarov is to be joined in space by Soyuz 2 for a rendezvous, but his craft develops serious control problems while in orbit and during an emergency descent under manual control, he enters the atmosphere at the wrong angle. The searing heat of re-entry partially welds his chute compartment shut, an emergency not solved until he is less than 5 miles from the earth—a distance too short for the chute to deploy properly. If Komarov is not already dead from the heat of re-entry, impact at 200 mph certainly kills him. On April 26, Komarov's ashes are taken to Red Square (**3**) for interment at the Kremlin wall. Cosmonaut Col. Georgy Beregovoi (**4**) will fly the next Soyuz mission, but not until October 26, 1968. Like Apollo, Soyuz is troubled.

COURTESY TASS FROM SOVFOTO

3

COURTESY TASS FROM SOVFOTO

COURTESY TASS FROM SOVFOTO

5

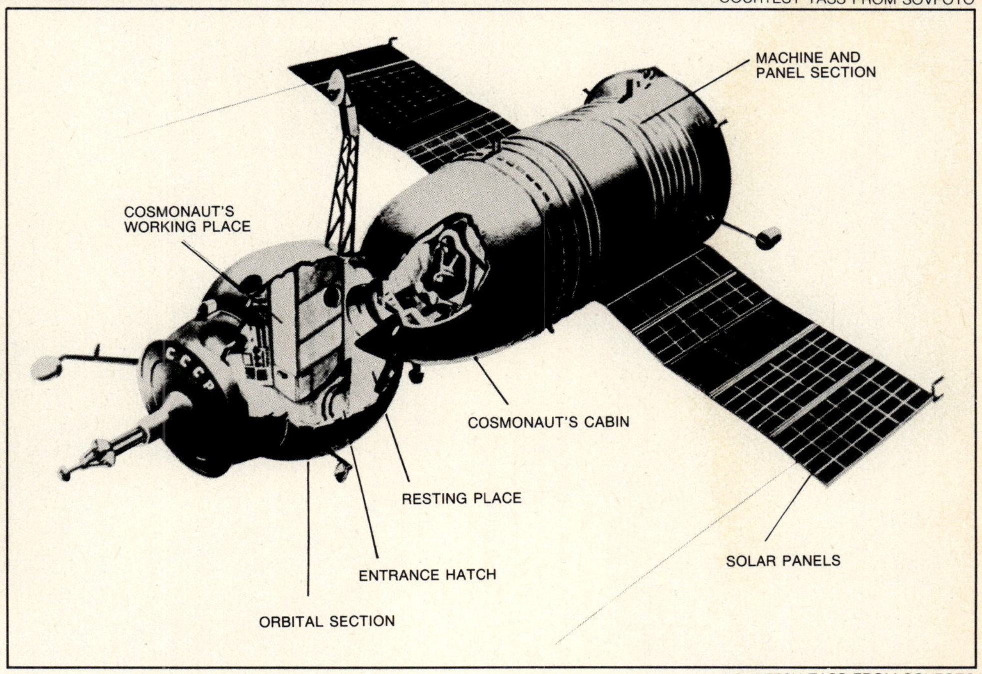

COURTESY TASS FROM SOVFOTO

SOYUZ, SALYUT & DISASTER

The flight of Soyuz 2 marks the Russian return from tragedy. Launched on October 25, 1968, it is an unmanned mission, but is joined in orbit the following day by Soyuz 3, piloted by cosmonaut Beregovoi. During his four days in space, Beregovoi makes two rendezvous with the target Soyuz 2, but does not attempt docking. After 64 orbits, he returns to earth, landing as scheduled on the snow-covered steppes outside Karaganda in Kazakhstan. Early in January 1969, the Russians decide to launch two Soyuz craft and attempt a manual docking in orbit. Soyuz 4 will be piloted (**1**) by cosmonaut Col. Vladimir Shatalov (3rd from left), while cosmonaut Colonels Evgeny Khrunov, Aleksei Yeliseyev and Boris Volynov (left to right) make up the crew of Soyuz 5. On January 14 at 10:39 AM Moscow time, Soyuz 4 lifts off (**3**) and the ground crew goes into action immediately to launch Soyuz 5 the next day. After Soyuz 5 achieves orbit, automatic approach systems bring the two spacecraft within 300 feet. Shatalov and Volynov then

1 COURTESY TASS FROM SOVFOTO

3 COURTESY OF SOVFOTO

take over manual control of their spacecrafts to attempt the docking. A camera aboard Soyuz 5 photographs the maneuvering of Soyuz 4 (**4**) as its "active" end moves toward the "passive"adapter cone. The two craft dock end-to-end, creating in effect an experimental space station with a crew of four. To earthbound observers, it is now abundantly clear that with this as a first step, Russia intends to master the technique of assembling a large space station—a permanent orbiting outpost to be supplied by shuttle vehicles from earth. The artist's rendition (**2**) previews the next event, or second phase of the combined Soyuz 4/5 mission, as cosmonauts Khrunov and Yeliseyev transfer from Soyuz 5 to Soyuz 4, moving slowly while grasping a special rail provided for the purpose, and retrieving the camera which filmed the docking as they make their way to the forward compartment of Soyuz 4. Spirits are high with the success of the rendezvous and docking maneuvers, both in the combined spacecraft and at Coordinating Data Processing Center on the ground. The Russian achievement is a substantial one; the two docked craft orbit as one for over four hours while the crew discovers that maneuvering, orbit corrections, and altitude control is not difficult. On January 17, the two Soyuz craft return to earth. Re-entry and landing are normal—Soyuz 4 carrying Shatalov touches down 25 miles northwest of Karaganda in Kazakhstan, and Soyuz 5 (**5**) lands safely approximately 124 miles southwest of Kustanai. Mission completed!

2 COURTESY OF SOVFOTO

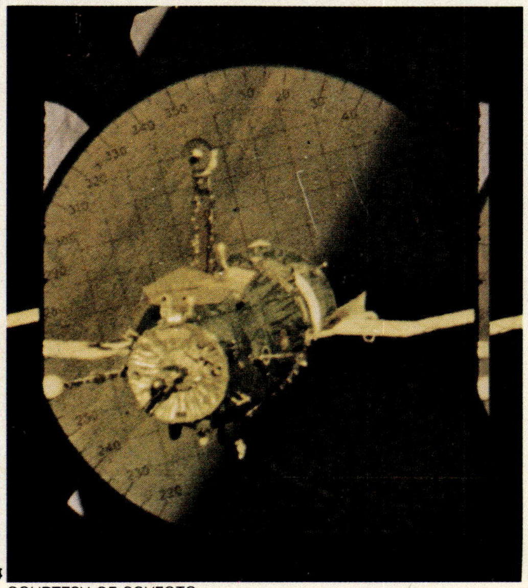

4 COURTESY OF SOVFOTO

5 COURTESY TASS FROM SOVFOTO

SOYUZ, SALYUT & DISASTER

Several months elapse with no Russian manned flight following the Soyuz 4/5 mission of January 1969. Subsequent Soviet space launches have been primarily concerned with the Lunik moon station program. The next Soyuz mission will begin in October, when seven cosmonauts (**1**) are scheduled to undertake the launch of three Soyuz spacecraft into orbit at 24 hour intervals. Georgi Shonin and Valery Kubasov (first row, left two) will fly in Soyuz 6; Victor Gorbatko, Anatoly Fillipchenko and Vladislav Volkov (left to right, back row) are to follow in Soyuz 7, and finally, Vladimir Shatalov and Aleksei Yeliseyev (first row, right two) will man Soyuz 8, flagship of this mission. As commander of the cosmonaut team, Shatalov (**2**) carefully checks the Zenith 80 cameras that will be used to record their venture. The Soviet cosmonauts intensify their training (**3**), with those new to space flight like Georgi Shonin (foreground) benefitting from the experience of Soyuz 4/5 crew members like Boris Volynov. By early

1 COURTESY TASS FROM SOVFOTO

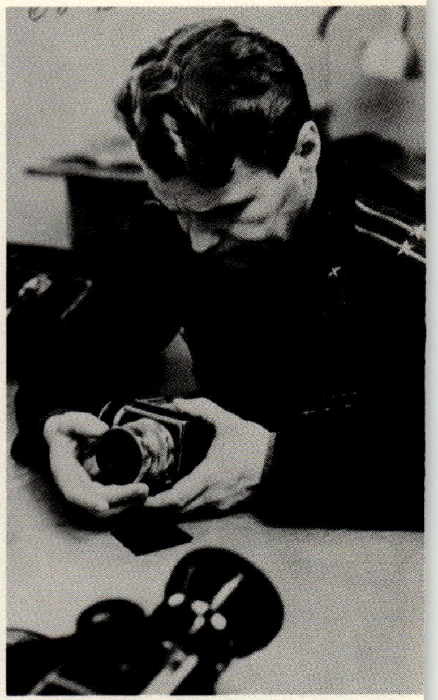

2 COURTESY TASS FROM SOVFOTO

5 COURTESY TASS FROM SOVFOTO

6 COURTESY OF SOVFOTO

October, all three crews have completed their training sessions in handling the Soyuz spacecraft, and arrive at the Baikonur Cosmodrome to await their departure. Meanwhile, the Soyuz 6 and its carrier rocket arrive on a platform truck and are hydraulically mounted in the launch tower. Once service masts rise (**4**) to form the service structure, the transportation platform holding the rocket swings away and down to a horizontal position. Soyuz 6 is fueled and checked out for launch. The group mission begins with the lift-off of Soyuz 6 at 2:10 PM Moscow time on a cold, rainy October 11; Soyuz 7 is launched at 1:45 PM the following day, and Soyuz 8 joins the two orbiting spacecraft with a blast-off at 1:29 PM on October 13. An amazed world fully expects the three Soyuz spacecraft to rendezvous and dock, forming a space station. No one in the west really knows what the Soyuz 6/7/8 mission plans actually are, but the multiple docking is not attempted, although the three spaceships do maneuver to within a few hundred yards of each other. On October 16, Valery Kubasov performs a series of welding experiments. At 12:52 PM that same day, Kubasov and Shonin land on the frozen steppes 110 miles northwest of Karaganda in a touchdown similar to that of Soyuz 9 (**5**) which follows some months later. At 12:26 PM the following day, Soyuz 7 lands near the same spot, and Soyuz 8 comes down 90 miles north of Karaganda on the 18th. The Soviet press greets them with expectations (**6**) of space stations.

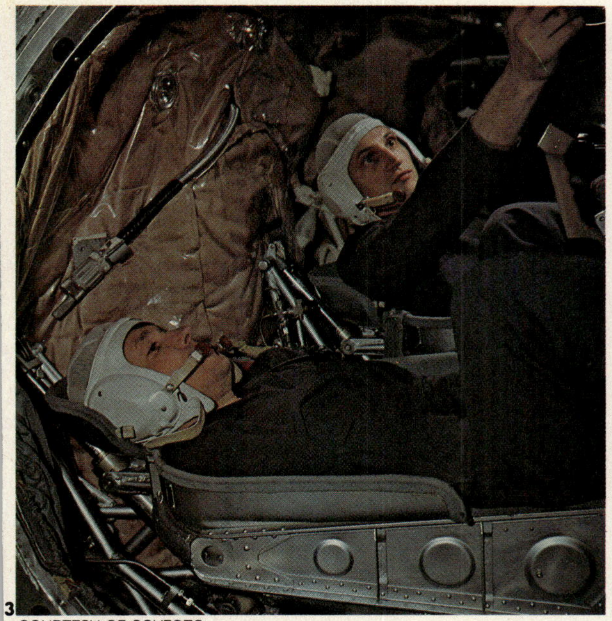

3

COURTESY OF SOVFOTO

4

COURTESY OF SOVFOTO

SOYUZ, SALYUT & DISASTER

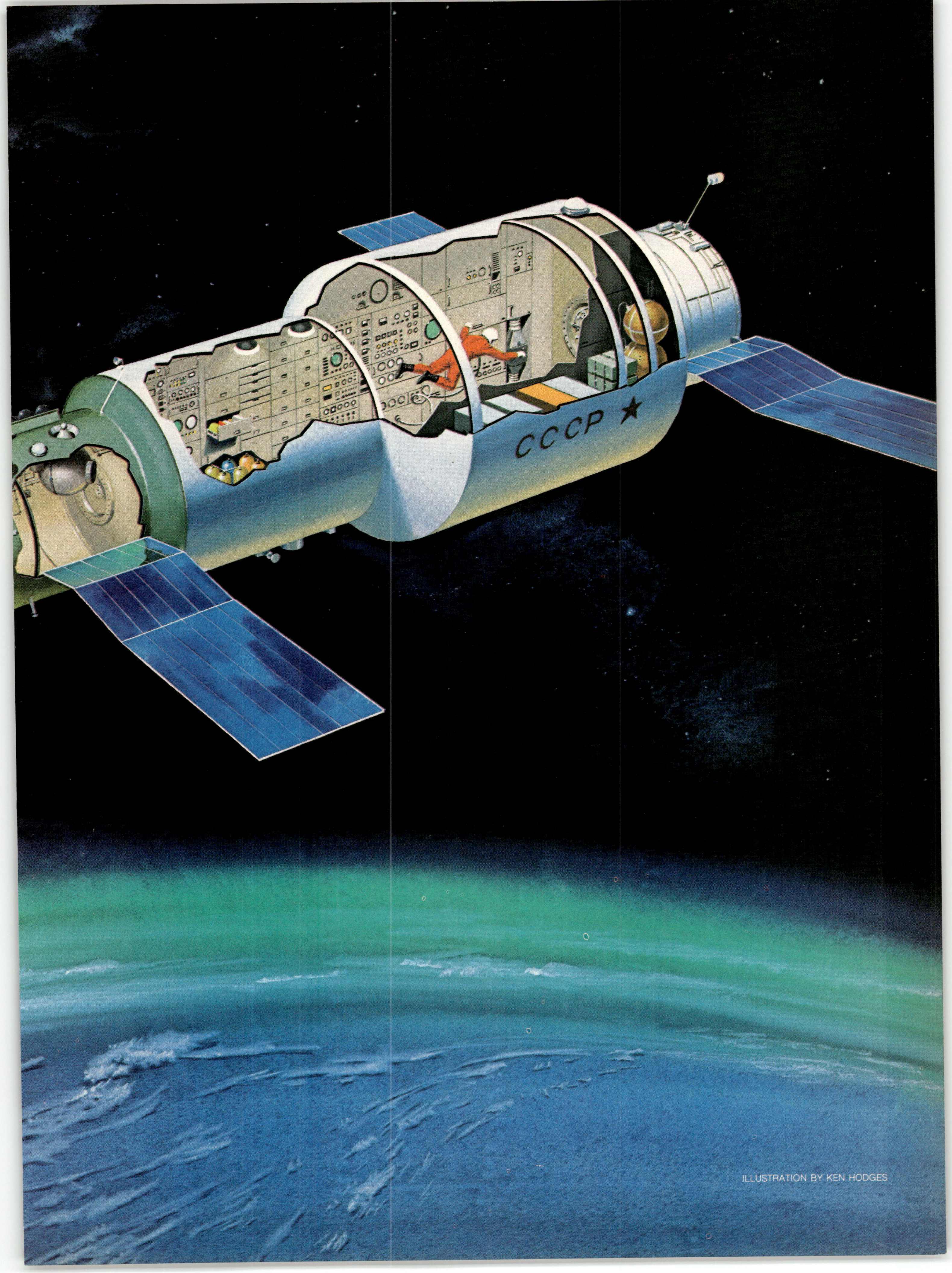

ILLUSTRATION BY KEN HODGES

SOYUZ, SALYUT & DISASTER

The 180-day mission of Soyuz 9 begins on June 1, 1970, with cosmonauts Andriyan Nikolayev and Vitaly Sevastyanov at the controls. The world waits expectantly, but Soyuz 9 is a routine flight to refine guidance, attitude control and stabilization systems. Ten months later (April 19, 1971) Russia launches an enormous space station named Salyut 1 (Salute), then prepares to launch Soyuz 10 and cosmonauts Vladimir Shatalov, Aleksei Yeliseyev and Nikolai Rukavishnikov to rendezvous on the 23rd (**1**). Soyuz 10 and Salyut 1 effect a rendezvous and docking as illustrated in cutaway on the preceding page (illustration based on photo of a model courtesy Novosti from Sovfoto). Although joined to Salyut for 5½ hours, no effort is made to enter the space station. Soyuz 10 and its crew return to earth on April 25. Soyuz 11 (**2**) undergoes launch preparation for a scheduled liftoff June 6, 1971. On the appointed date Georgy Dobrovolsky, Vladislav Volkov and Viktor Patsayev blast off to a rendezvous with Salyut 1

1
PHOTO COURTESY OF SOVFOTO

2
PHOTO COURTESY OF SOVFOTO

and the manning of the world's first orbiting space station. The cosmonauts spend 22 days aboard Salyut 1 conducting various scientific experiments and physiological tests. They are in good spirits as they reactivate their Soyuz spacecraft and undock at 9:28 PM on June 29. A few hours later (1:35 AM June 30), Soyuz 11 begins re-entry, and with the exception of a communication loss as the braking rockets are fired, its return to earth seems otherwise normal. Moments after landing, the search helicopters arrive and find no one outside the craft. Opening the cabin hatch, Dobrovolsky, Volkov and Patsayev are found still strapped in place—*dead.* For the second time, tragedy has struck the Soyuz program. A leak had apparently developed in the cabin about 30 minutes before landing, causing a loss of pressure and instant death for the crew, as Soyuz cosmonauts do not wear space suits. Two days later (July 2, 1971) Soviet leaders Brezhnev (left) and Kosygin (right) lead the funeral procession (**3**) bearing the cosmonauts' ashes through Red Square to the Kremlin Wall where they are interred near those of Vladimir Komarov and Yuri Gagarin's. Gagarin perished in a training plane crash March 27, 1968. The death of the Soyuz 11 crew comes just a few months after the Russians' Luna 17 automatic space station deploys Lunokhod 1 (**4**) on the moon. This six-foot trolleylike vehicle is propelled by batteries powered by the sun. Lunokhod 1 is equipped with remote control cameras and carries out many experiments.

3 PHOTO COURTESY OF SOVFOTO

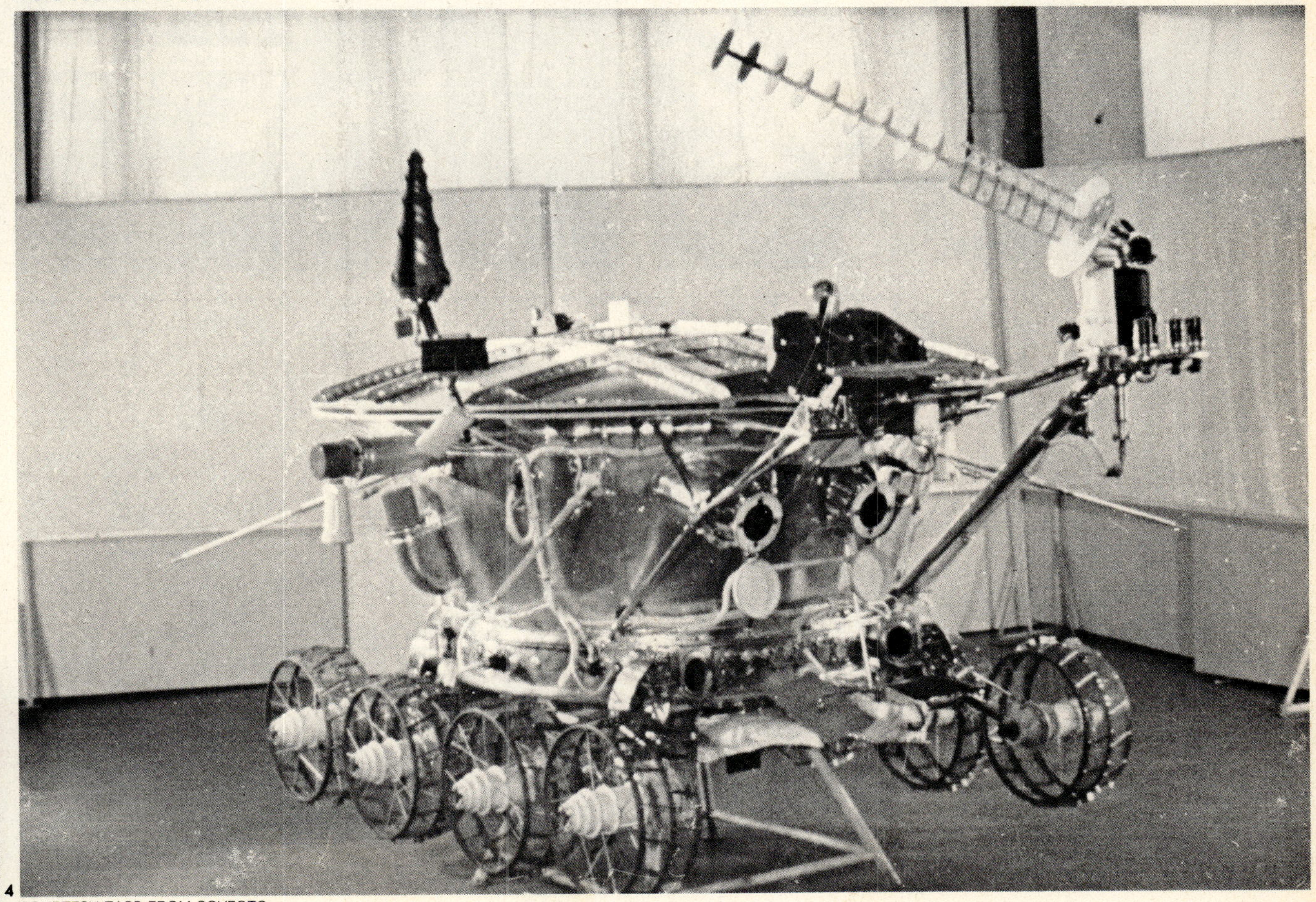

4 COURTESY TASS FROM SOVFOTO

To send Americans to the moon will require the development and launching of a rocket seven times more powerful than any in existance. Sixteen days after John F. Kennedy takes office as President of the United States, three of the nation's top rocketry experts meet to decide where this will be done. Merritt Island next to Cape Canaveral is selected to avoid costly duplication of facilities and systems needed to create the first operational spaceport (**1**), which will gain fame as Complex 39. Land acquisition begins March 1, 1962 with the outright purchase of 83,894 acres (**2**) and will eventually cost over 71 million dollars. The Space Center's southern boundary is to follow the Barge Canal connecting Port Canaveral with the Indian and Banana Rivers and parallel the southern tip of the Cape. Sufficient space is acquired to accommodate future projects such as lunar vehicles, a space shuttle and even nuclear-powered rocket launches. Provision is made by NASA to preserve the ecology of Merritt Island and

KSC & COMPLEX 39

Florida swampland is transformed into a staging area for a lunar journey

1

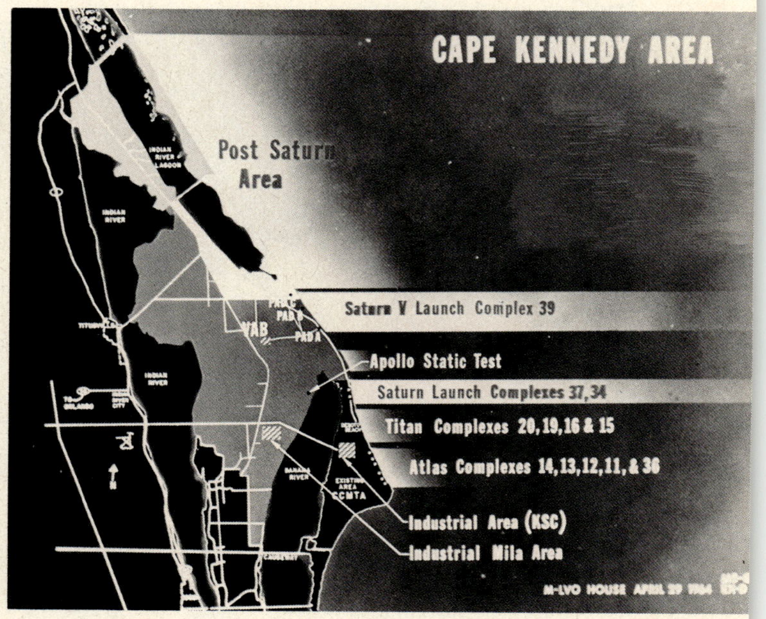

2

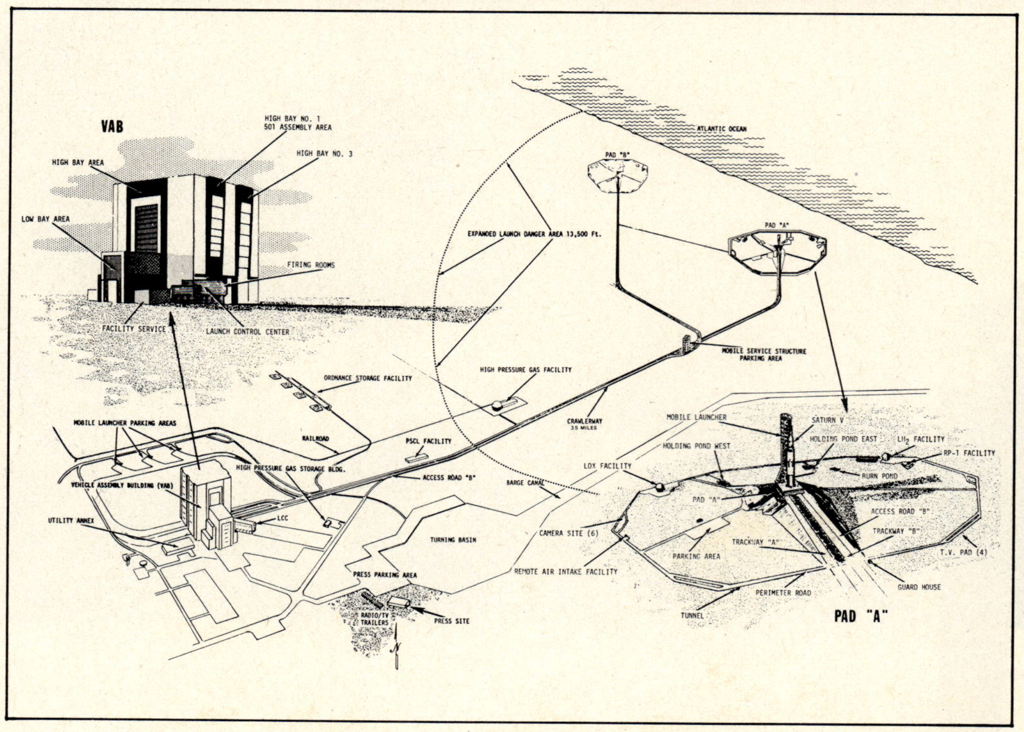

3

surrounding areas. In fact, surely one of the more amazing facts about the spaceport is that the whole area is a wildlife sanctuary. A change in the concept of launch preparation for the giant Apollo/Saturn V dictates the physical layout (**3**) of Complex 39. Standing 363 feet high, the Saturn's three stages will be assembled on a movable launch platform inside a special hanger, the Vehicle Assembly Building (VAB). A 3.5-mile road or crawlerway will connect the VAB with the two launch pads. Once the Saturn V is in place on the pad, a Mobile Service Structure will permit launch preparations. All preparation activities are monitored from the Launch Control Center (LCC), located beside the VAB. A construction task quite unlike any previously attempted, Complex 39 is labeled by *Architectural Forum* magazine as "one of the most awesome construction jobs ever attempted by earthbound men." While this takes form and shape, NASA directs the construction of a lock at Port Canaveral to connect the Banana River with the Atlantic Ocean (**4**). As two of the Saturn's three stages must be transported by water due to their size, this will allow ocean-going barges direct access to the Space Center once a channel is dredged in the Banana River. Saturn V's takeoff point will be a concrete hardstand, or pad (**5**) covering one-half square mile. After the death of President Kennedy from an assassin's bullet November 22, 1963, the spaceport he championed was renamed in his honor the John F. Kennedy Space Center, Cape Kennedy, Florida.

4

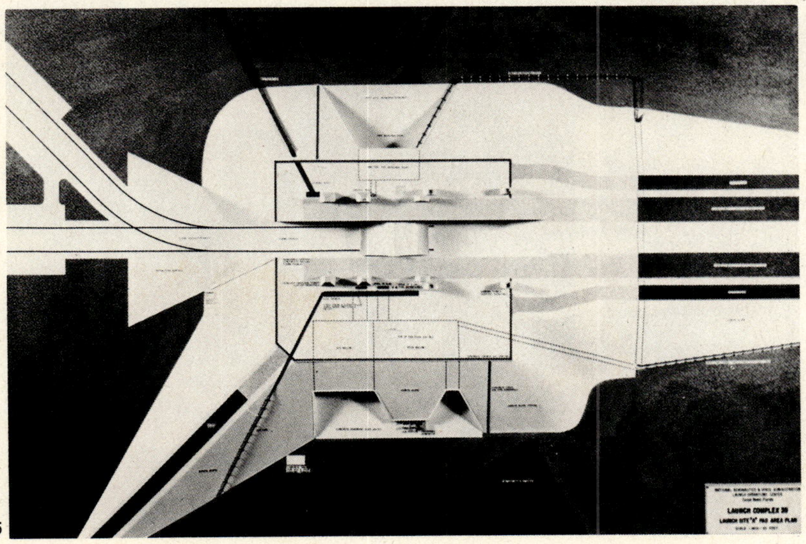

5

KSC & COMPLEX 39

Construction begins in July 1963 on the heaviest portable structure ever built by man, the 10.5-million-pound mobile Launcher and Umbilical Tower. With its 380-foot-tall tower (**1**) permanently installed on a two-story, 18,000-square-foot base, the entire mobile launcher (LUT) extends 445 feet into the air and serves as an assembly platform for the Apollo/Saturn V. Once the rocket's three stages are in place, the entire LUT is transported to the pad where it serves as a launch platform. As the steel skeleton of the first LUT takes shape, other necessary tasks are underway. Launch pads that resemble ancient Mayan pyramids rise out of the swampland. The initial 80-foot pile of fill is left to settle for four months, during which it causes the sub-soil to compress four feet. The fill is then levelled to form a hardstand (**2**) 48 feet high with its approach ramp (behind the pad, as seen here) and a flame trench of reinforced concrete running through the middle. Altogether, this site must be capable of supporting approximately

2

1

3

18,000 tons of weight. Moving the Saturn V and LUT to the pad is to be accomplished by a Crawler/Transporter (**3**). After considering several different possibilities, NASA settles on an adaptation of the Marion coal strip-mining shovel that is completed in 1962. This giant machine incorporates proven tank-track principles for movement and hydraulic jacking accurate enough to maintain a constant platform level. This is the most important link in the chain—if the mobile concept is to work, the assembled Saturn V and its launch platform must be moved from the VAB to the pad with an absolute minimum of vibration, as thousands of mechanical and electrical connections must make the trip without damage. The Crawler/Transporter will also be used to move the Mobile Service Structure (**4**) into position to provide connection, checkout and fueling facilities for the Apollo/Saturn V during its final phases on the launch pad. Originally designed as a fixed installation at Pad A, this derrick-type, steel-trussed tower is changed to a mobile design in mid-1962 and fabrication begins in September 1964. By late 1962, coordination of planning data from various manufacturers has advanced enough to begin designing the VAB. After considering various shapes, a back-to-back box-like structure is selected and 58,000 tons of steel begin to rise into the air (**5**) atop 4,225 steel pilings. By the end of 1964, the structural framework is completed—over a million bolts hold the 45,000 pieces of steel in place. Complex 39 takes on form.

5

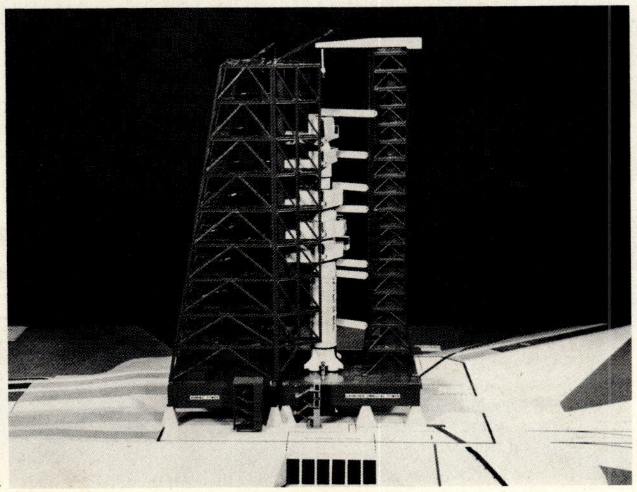

KSC & COMPLEX 39

Things have changed considerably in the years following the first rocket launch from Cape Canaveral (July 24, 1950), especially in the control systems used. From a small concrete/steel blockhouse and a tarpaper shack, control centers grow more complex as the state-of-the-art advances. But while Cape Canaveral's blast-proof blockhouses remain a few hundred feet from its launch pads, the eight-story tall Launch Control Center (LCC) at Complex 39 (**1**) is a huge four-level building southeast of the VAB, and connected to it by a corridor six floors from the ground. The first level houses administrative offices and support facilities, while the second one is devoted entirely to telemetry, instrumentation and other data transmission/evaluation equipment. Final countdown and launch of the Apollo/Saturn V will be conducted from one of the three firing rooms (**2**) located on the third level. This is the heart of the LCC, containing the computers that form the electronic "brain" which controls inspection of all three stages while un-

1

2

der assembly "next door," as well as the actual launch operation. Looking down on the firing rooms, the top level contains four rear-view projection screens per firing room. These allow personnel manning the 470 monitors and control consoles to watch four different views of the launch site during an operation. The LCC also houses the Range Safety Officer (**3**), whose little-publicized role is crucial to a launch. Should an emergency arise and the mission have to be aborted, he is responsible for activating the propellent dispersion system built into each stage of the Saturn V. When he throws a switch, the engines will shut down and an explosive device rips open the fuel tanks, emptying the rocket of propellant before it is destroyed. If a serious malfunction occurs during a countdown without sufficient time to leave the launch area, the Apollo Emergency Ingress/Egress and Escape System is used. This provides access to and from the Apollo Command Module, as well as an escape route and safe quarters for 20 people. Part of this system is the escape tube (**4**) being tested by astronaut Charles Duke. This leads from the top level of the mobile Launcher Umbilical Tower (LUT) to a Blast Room below mounted on coil springs to absorb explosive forces. The astronauts can reach the launcher base by elevator in 30 seconds, or take a cable cage (**5**) to a bunker one-half mile from the firing site. The specially-designed transporter van (**6**) stands ready for an emergency until the area is cleared as the countdown nears an end.

3

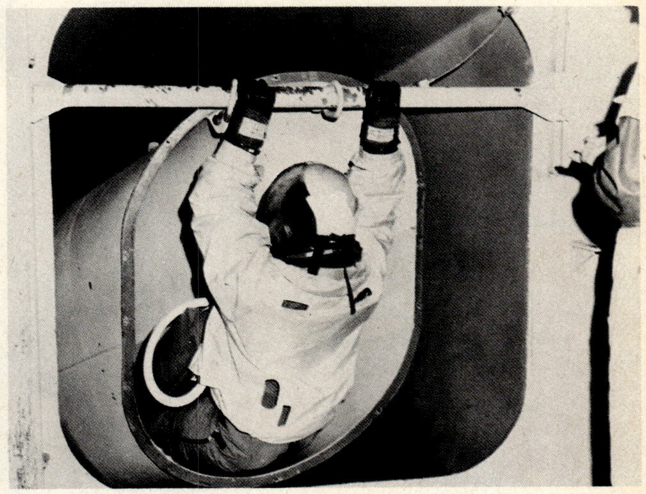

4

5

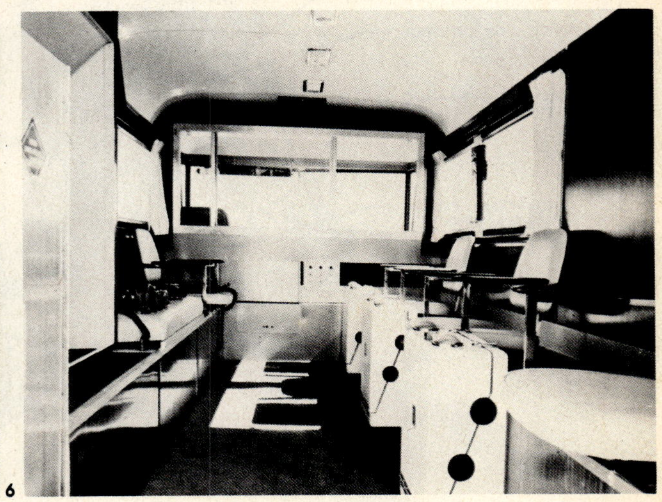

6

KSC & COMPLEX 39

Over two years (1963-65) are required to construct the three Launcher/Umbilical Towers for Complex 39. These provide one for each of the VAB's high bays and allow preparation of up to three Saturn V launch vehicles at the same time. When not in use in the VAB or on the launch pad, they are stored (**1**) in a parking area just to the north of the VAB. Regardless of where they are located, each LUT rests on six pedestals approximately 22 feet high; this "headroom" between ground and LUT is necessary to allow the Crawler/Transporter to drive under the LUT to move it. Three feet in diameter, each of the six pedestals can be individually adjusted up to four inches in height. When the LUT is ready to be unloaded from the Crawler/Transporter, huge bolt collars on each pedestal are rotated until all are of the correct height and level with the LUT. The platform is then connected to each pedestal by 12 high-strength bolts. When the LUT is positioned at the launch pad (**2**), a second support system of four more columns (one

1

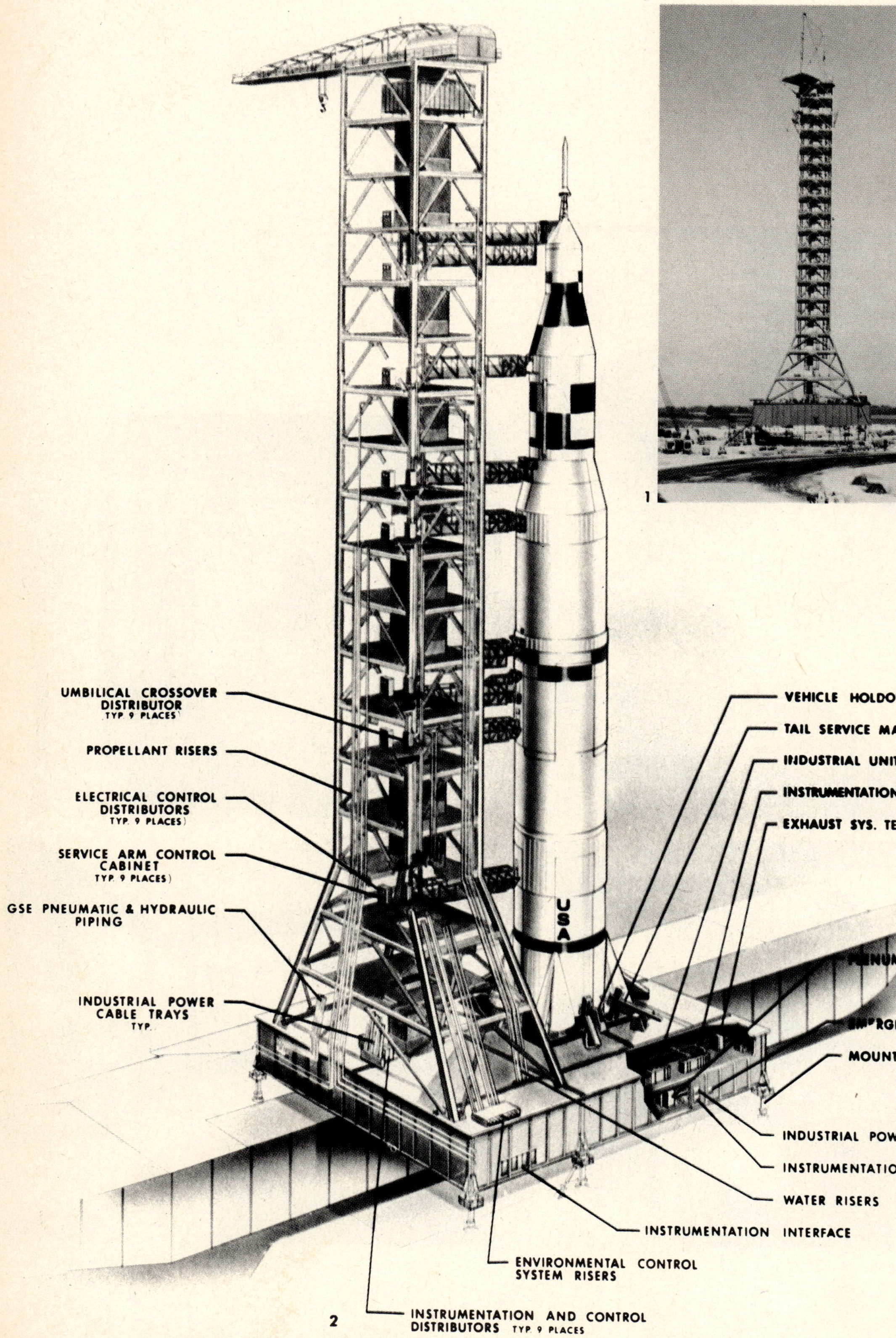

2

at each corner of the engine exhaust opening) is used to make certain the launch platform is sufficiently rigid for blast-off. The two-level launcher base is equipped with hold-down arms that keep the Saturn V in place for 8.9 seconds until the computer verifies that each engine is at full thrust. The base also contains equipment and computer rooms, the floors of which are suspended on springs while their walls are lined with two inches of thermal/acoustical insulation. This reduces the interior noise level during a launch to less than that created by a 747 jet. Permanently positioned on the launcher base, the 380-foot high umbilical tower provides support for the nine swing arms and 17 work platforms required to hold and service the Saturn V. This umbilical tower is capped by a 25-ton, remote-controlled, hammer-head crane that can reach to the bottom of the flame trench if necessary. Special doors in the VAB are required to permit entry and exit (**3**) of the LUT from the high bays. Shaped like an inverted T, the lower door consists of four leaves that slide open horizontally, while the upper section uses seven leaves that telescope open vertically, one behind the other. Opening this door system requires 45 minutes and once the LUT has passed through (**4**), another 45 minutes to close. A better idea of the mobile launcher's enormous size can be gained by the scale provided (**5**) when automobiles are parked beside them in the storage area. In 1965, many doubted the mobile concept would work; the LUTs are ready by 1968.

3

4

5

KSC & COMPLEX 39

Activity at Complex 39 centers around the Vehicle Assembly Building (VAB), where the Saturn V is to arrive in stages from test sites for assembly and checkout. Construction begins in November 1962 with clearing and filling operations. To raise the site level by seven feet, 1½ million cubic yards of soil are dredged from the Banana River. The building design selected (**1**) is that of two rectangular boxes joined together—a High Bay (524 feet high) and a Low Bay (210 feet high). Each stage will be checked in the Low Bay and then transferred by overhead crane to the High Bay for assembly. Because of the hurricane-force winds that sweep the Florida coast, a foundation of 4,225 steel pipe pilings each 160 feet long is installed before pouring the concrete floor slab. This superstructure will prevent the VAB from blowing apart during strong winds. A chemical salt solution found below the surface has to be neutralized to prevent foundation corrosion or electrolysis will turn the VAB into the world's largest wet-cell bat-

1

3

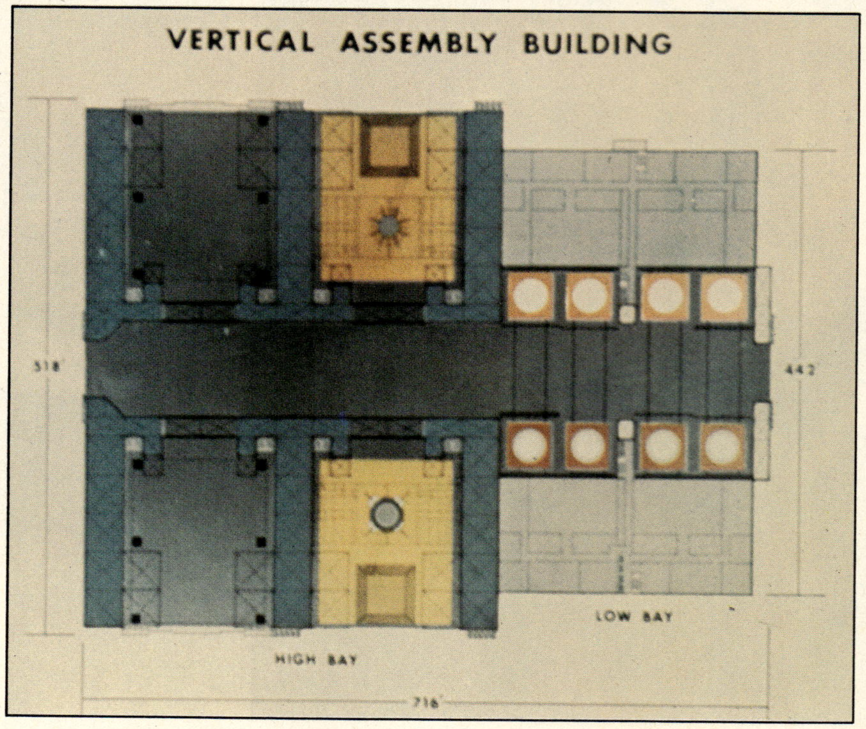

2

tery. Mobile work platforms are designed in the form of self-contained buildings that are air-conditioned and dustproof. These can move both horizontally and vertically to enclose the Saturn V as it is assembled. When not in use, they are stored against the VAB walls. The four High Bays are located in back-to-back pairs (**2**) with a single transfer aisle serving all four. A similar Low Bay design allows technicians to work on as many as eight upper stages at the same time, two in each Bay. Constant circulation of the air inside is required to prevent formation of clouds and artificial rain in the VAB. Some 18 months after erection begins, the VAB is complete. Although one of the world's largest buildings, the massive structure facing Complex 39 (**3**) is not visually impressive from its outside, as there are no other buildings in the vicinity. Not until one enters the VAB can he fully appreciate the incredible size of the building, which can lean as much as 12 inches during high winds without structural damage. The Low Bay section (**5**) contains eight preparation and checkout "cells" for the Saturn's upper stages, each equipped with work platforms that surround the stage. Living quarters for flight crews and support technicians are provided in the Low Bay area for use during critical work periods. Late in May 1966, the doors of Bay 1 are opened and the transporter emerges with a Mobile Launcher carrying the AS-500F, an exact duplicate of the Saturn V used for training purposes. Rollout (**4**) begins the 3.5-mile trip to the launch pad "A".

4

5

COPYRIGHT 1973 BY CTEIN

KSC & COMPLEX 39

While construction of the Vehicle Assembly Building, Launch Control Center, Launcher/Umbilical Towers and Launch Pads is underway, work begins on the Crawler/Transporter which will carry the LUT from the VAB to Pad A or B for launch. Once the tank-track design is decided upon, contracts are let and early in 1963, the Marion Power Shovel Company begins work on components in its Marion, Ohio plant. Subassembly is done there, with completed sections shipped to Complex 39 where final assembly (**1**) takes place. The dimensions of the Transporter (131 x 114 feet) are in scale with the gigantic proportions of the other facilities at Complex 39; the platform or chassis which carries the LUT will accomodate 23 men shoulder-to-shoulder (**2**) across its width. In order to carry the ML perfectly level at all times despite the terrain and the five degree ramp climb, two independent hydraulic servo systems are installed. Each has a manometer and transducer to sense errors in leveling. These send signals to the servo system,

1

2

which hydraulically operates the 16 leveling cylinders. A double-tracked crawler truck supports each corner of the chassis (**3**). These can be steered in pairs at either end, but as the operator cannot see behind or to opposite sides of his cab, ground observers equipped with walkie-talkie radios are required to walk beside the Transporter and give instructions. The entire Apollo/Saturn program has been based on the concept of mobility—a departure from rocket assembly on the launch pad—and this mammoth moving machine is the key element in its success. Two transporters are built at a cost of about 14½ million, with final "bugs" eliminated in July 1965. The Crawler/Transporter will also be used to move the Mobile Service Structure (MSS) from its parking area (**4**) some 7000 feet from Pad A. Like everything else at the Space Center, the MSS is extraordinary—a steel-trussed tower standing 402 feet and weighing 5000 tons. The MSS is positioned by the Transporter to service the Apollo launch vehicle and spacecraft on the pad. Five work-level platforms provide 360 degree access and can be positioned as desired. Some seven hours before launch, the Transporter returns the MSS to its parking area. To dissipate flames and minimize damage to the LUT and pad during a launch (**5**), a water deluge system will pump 50,000 gallons of water a minute into the flame trench while the rocket engines build up to full thrust. Despite this, the flame deflector will turn to glass and about 1½ inches will be burned off the launch pad.

3

4

5

The voyage to the moon represents an undertaking on a scale never before attempted. Not only has technological capability been exploited to its fullest, so have scientific and management methods as well. Much of the effort and attention is focused on the huge and powerful Saturn V launch vehicle. It requires anywhere from 42 to 45 months to complete one of these gleaming white and black giants that will soon take man on his most incredible voyage. The elements surrounding the creation of the Saturn V are as nationwide (**1**) as the logistics that bring the widely scattered components—some six million of them—to one point: Cape Kennedy. All these components and sub-assemblies, constructed as far away from the final launch site as Seattle, Washington, or as close as West Palm Beach, Florida, must come together at the right time in the bays of the Vehicle Assembly Building (VAB) (**3**). In the Saturn V's case, an entirely new method of assembly and test had to be developed to thwart Florida's weird weather

ASSEMBLING THE SATURN V

Six million pieces must come together in the right place at the right time

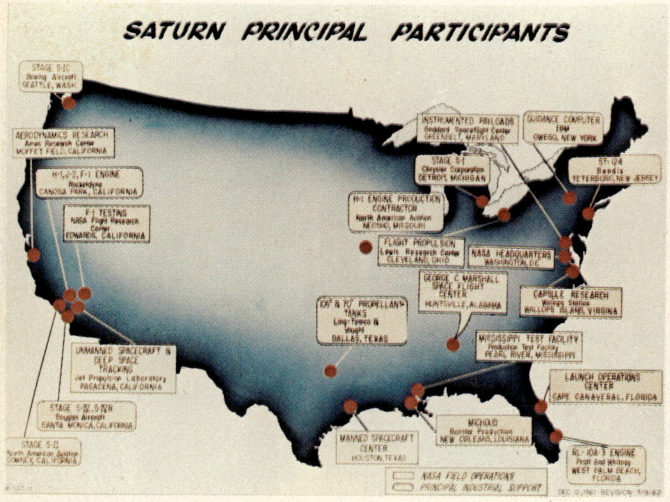

1

2

3

and the corrosive effects of the salty ocean spray. It is too time-consuming and expensive to take a Saturn V apart to replace one faulty piece, so every part is tested separately and in combination with each part that's added as assembly progresses. How is the Saturn V assembled? Much like a sandwich, with everything stacked in proper order and the small items added as they are needed, with constant testing to insure perfection. Before assembly, however, must come construction and transportation. The engine shrouds and tail fins for the first-stage (S-1C) booster are constructed at Michoud, Louisiana (**2**). Far away in Huntington Beach, California, the S-IVB third stages take form in the workshops of Douglas Aircraft (**4**). In the foreground is the forward section, while in the background is the tapering interstage structure that mates the 24-foot-diameter third stage to the 33-foot-diameter second stage (S-II). New welding techniques had to be developed for the Saturn's thin-skinned structure. The 33-foot-diameter lower stage structure requires electric welds accurate to plus-or minus .013-inch over the 100-foot circumference. At the TRW facility in Redondo Beach, California, the descent engine (**5**) of the Lunar Module undergoes one more inspection prior to shipment to Grumman in New York. Also destined for Grumman's Bethpage plant is the LM's ascent stage engine (**6**) being given final touches by employees of Rocketdyne in Canoga Park, California. Rocketdyne builds most of the engines on Saturn V.

4 COURTESY McDONNELL DOUGLAS

5 COURTESY TRW

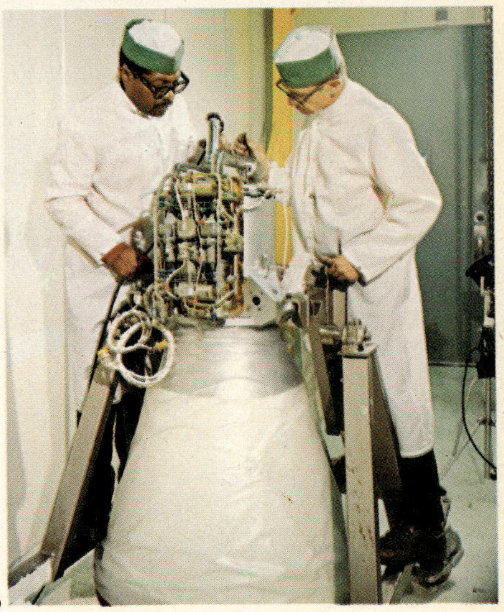

6 COURTESY ROCKWELL INTERNATIONAL

ASSEMBLING THE SATURN V

Once the individual contractors have completed the various components, each must be transported to the Cape, or to one of NASA's special testing facilities for final checkout. Considering the immense size of most hardware that makes up the Saturn V/Apollo Spacecraft, coupled with their delicacy, shipping becomes a story in itself. Before leaving the contractor's, components are checked and then packed to protect against shock, weather and ingress of foreign material. A view of an S-IVB third stage (**1**) in the VAB, shows it mounted on its own trailer. Hoops encircle the full diameter of the outer skin at either end. Contoured frames on the trailer mount rollers which allow the stage to be rotated when the clamps securing the hoops to the trailer frame are unlocked. A dust-tight cap covers the engine nozzle. Ascent stages of the Lunar Module are packed in sealed containers (**2**) and sealed with inert gas. They are shipped aboard one of Aero Spaceline's "Super Guppy" transport aircraft. The "Sup-

1

2

er Guppy" also ships the S-IVB third stage in its outsized airframe. Moving the immense first and second stages is undertaken by NASA via our waterways (**3**). From their assembly point, Michoud, Louisiana, the stages are carried by river barge 1086 miles north to Marshall Space Flight Center in Huntsville, Alabama. There, at the immense Redstone Arsenal, they are given a thorough checkout and static firing to test the integration of all systems. In order to reach the test site, the first stage is hauled by special trailer (**4**). The S1-C first stage is a cumbersome piece of hardware to move on land, even without its engine shrouds and fins! One of the launch vehicle's most critical systems is the Instrument unit (IU), which mounts atop the third stage. Built by IBM's Federal Systems Division in Huntsville, the completed IU is shipped by Aero Spacelines (**5**) to the Cape where it receives its final checkout. Shrouded, and rigidly supported, the IU is Saturn's brain. After completion of testing at Redstone Arsenal, the Saturn components are loaded back onto the NASA barges and shipped back to the mouth of the Mississippi River, from where they make the 932.4-mile sea trip through the Gulf and around the Florida Peninsula to Cape Kennedy. Here (**6**) the NASA barges "Poseidon" and "Promise" pass the New Orleans skyline. The "Poseidon" on the left is carrying the first Saturn V booster, while the "Promise" is carrying a Saturn IB first stage. Later Saturn V boosters will be static fired at NASA's test facility in Hancock County, Mississippi.

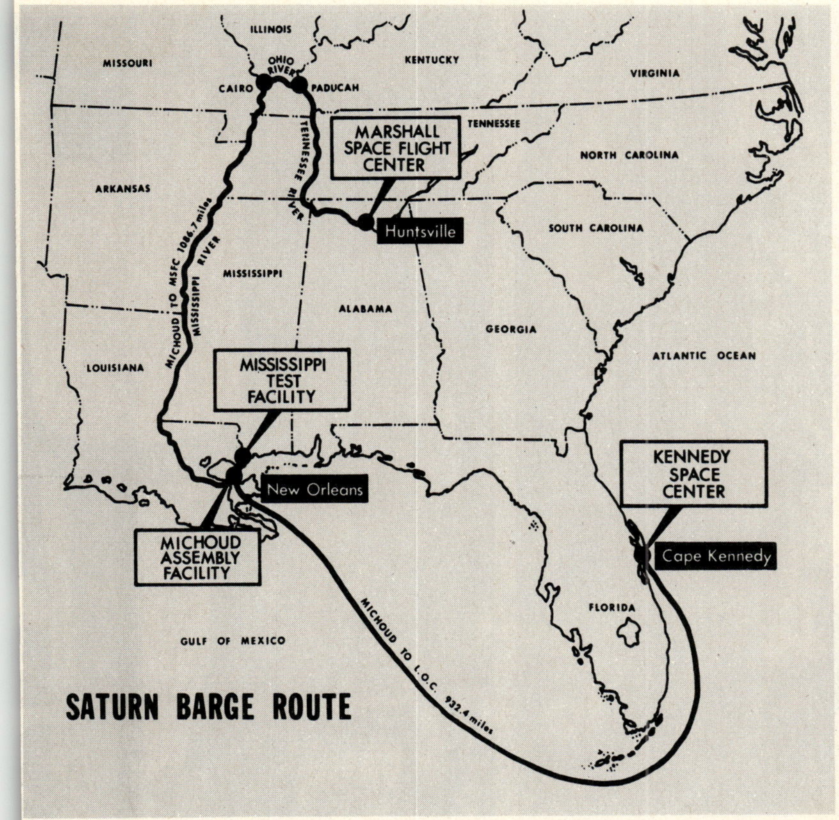

4

6

COURTESY AERO SPACELINES

ASSEMBLING THE SATURN V

Before stacking a Saturn V, the Crawler/Transporter moves one of the three Launch/Umbilical Towers (LUT) into a high bay in the Vehicle Assembly Building (VAB). Then the first stage is unloaded from its barge at the turning basin and transported via special dolly into the VAB transfer aisle (**1**) where it is hoisted upright by a 250-ton bridge crane that is one of two such heavyweights that straddle the high bays. The crane operator then lifts the S-IC (first stage) over 190 feet in the air (**2**) and moves into high bay where he carefully lowers it onto the four Hold Down/Support Arms at the edge of the 45-foot-square exhaust chamber opening in the LUT. The entire weight of the Apollo/Saturn V rests upon these arms. Once the S-IC is in position on the mobile launcher, the previously-built engine shrouds and fins (**3**) are attached along with the F-1 nozzle extensions and connections for the three Tail Service Masts. The second, or S-II, stage (**4**) comes next. It is the same diameter as the

1

2

3

303,000-pound, 138-foot-long first stage, but is only 81½-feet in length. While the S-IC is being erected, the S-II, S-IVB (third stage) and Instrument Unit (IU) are being inspected in the low bay checkout cells. The 72,000-pound S-II stage is removed from the low bay by crane and moved to the transfer aisle (**5**) where the 250-ton bridge crane is called upon to lift it up and stack it atop the first stage (**6**) with the interstage shroud concealing the five J-2 rocket engines. The shroud mates to the attachment skirt of the first stage with a precision that requires the crane operator be able to detect movement of 1/100-inch or less and be able to move the stage in any direction as slowly as 1¼-inches per minute. Initial attachment of the S-II second stage is made while it is still suspended from the crane. After mating, the retracting service floors roll together, surrounding the stages. The floors are mounted on the sides of the bay and can be adjusted in or out, up or down. Offices in the high bay are occupied by contractors and are convenient to service floors enclosing that stage. Thus Boeing's offices are next to the first stage, North American Rockwell's the second stage, etc. Immediately after erection of the first stage, the Instrument Unit is moved out of the checkout cell in low bay and placed on a platform in the high bay area near the S-IC where a hookup and checkout of the first stage and Instrument Unit is performed. After stacking the second stage, it, too, is tested by itself and with the other stages to ensure perfect operation.

4

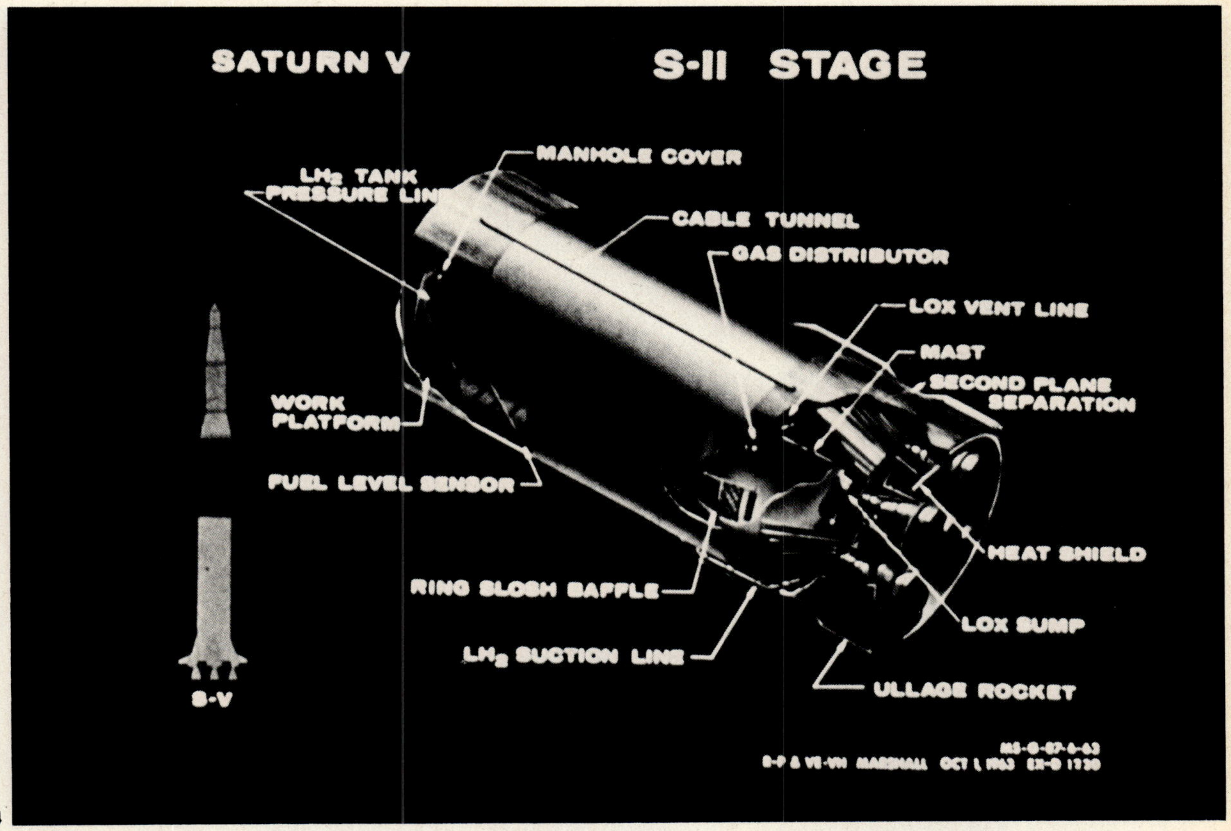

5

6

ASSEMBLING THE SATURN V

After the first and second stages are stacked, the third, or S-IVB stage is raised into position in the High Bay area. The S-IVB stage (**1**) is not new; it is a proven item of launch hardware from the Saturn 1B. However, its J-2 engine has been uprated to have a restart capability when used with the Saturn V lower stages. During a lunar mission, the J-2 engine will be burned twice, once after second stage burnout to place the Apollo Spacecraft into earth orbit. The second firing, of much longer duration, will push the spacecraft out of earth orbit and into a lunar trajectory. Both the rendering and photos on this spread show the S-IVB with its tapering attachment skirt in place. This skirt encloses the engine and takes up the reduction in diameter between the second stage at 33 feet and the S-IVB at 22 feet. The third stage is 59 feet in length and weighs 33,600 pounds empty of fuel. Similar to the S-II stage, the S-IVB engine burns liquid hydrogen and liquid oxygen, carrying a total of 230,000 pounds of fuel. Launch

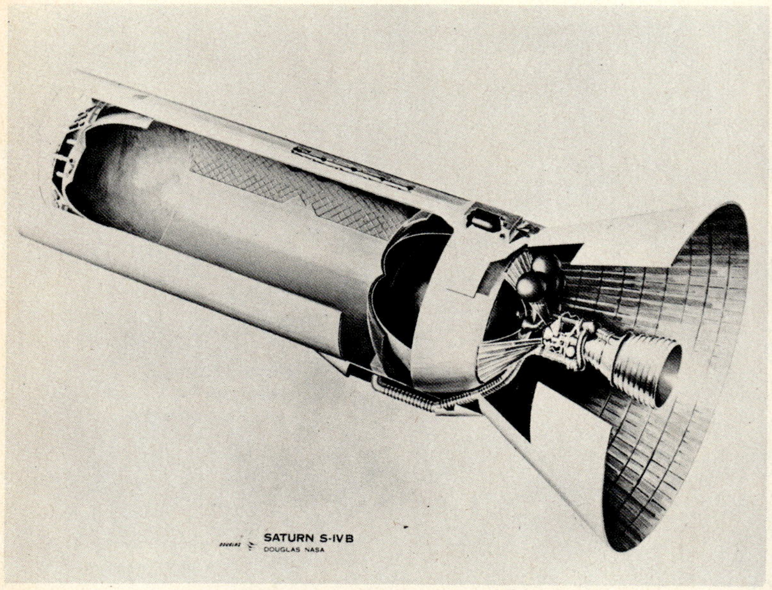

1

2

3

weight of the S-IVB is about 265,600 pounds. Up, up, up goes the third stage (**2**) for its rendezvous with the second stage skirt (**3**) high in the VAB. Note that the protective cover on the J-2 engine remains in place until just prior to mating of the two interfaces. The final step in completing the stack is the installation of the Instrument Unit (**4**). Smallest of the Saturn components, it is nonetheless one of the most critical, because it contains all the electronics for the launch vehicle's guidance, control and telemetry systems. Weighing 4,500 pounds, the Instrument Unit stands only three feet high and is 21 feet eight inches in diameter. The Instrument Unit has a further capacity to fulfill in a lunar mission, that of checkout and guidance control of the S-IVB third stage during its injection firing that starts the spacecraft on the way to the moon. After installation, IBM technicians complete the wiring and checkout of the ring (**5**). When installation of the instrument ring has been completed, the launch vehicle stands 281 feet high. The Apollo spacecraft, partly consisting of the Command and Service Modules (**6**) is assembled and tested as a unit separately from the Saturn V launch vehicle. When fully stacked and ready for launch, the Saturn/Apollo will stand 60 feet taller than the Statue of Liberty, and weigh 13 times more! The fuel capacity alone is enough to drive an average automobile 18 million miles, were the fuel to be gasoline. But the voracious appetite of the Saturn will devour it all in just a few minutes.

4

5

6

ASSEMBLING THE SATURN V

During launch and earth orbit, the Lunar Module is safely enclosed within a special housing called the Spacecraft/Lunar Module adapter, or SLA. The SLA, with LM inside, sits beneath the Command and Service Modules. The SLA consists of two main parts, a lower tublike section (**1**) into which the LM is lowered during assembly. The LM rests on special mounting pads around the circumference of the lower SLA section. The upper section of the SLA consists of four panels (**5**) which surround the upper half of the LM. These panels are hinged at the base and open up like petals on a flower, enabling the Command and Service Modules to rotate 180 degrees and dock with the LM, pulling it out of the lower SLA section. The SLA panels are heavily insulated to protect the LM during launch from heat generated by atmospheric friction. After the Command and Service Modules are assembled, and the upper SLA panels fitted together, the completed spacecraft is hoisted into the high bay of the VAB (**2**). The retracted legs

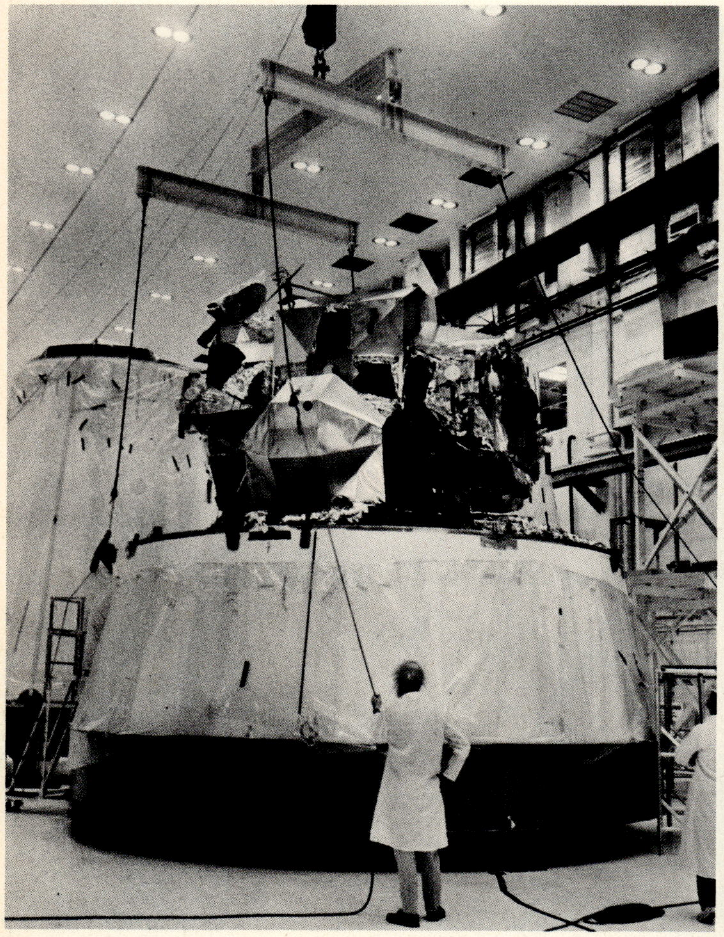

1

2

of the LM are visible beneath the SLA section. Finally, the spacecraft is mounted atop the launch vehicle (**3**), and one of the most miraculous achievements of modern engineering is nearly complete. The vehicle subsequently undergoes a series of tests, ensuring compatibility or that correct mechanical and electrical connections have been made throughout. After VAB testing is concluded, the prescheduled rollout takes place, and the Saturn V is moved to one of the two launch pads. The "capper", so to speak, on the Saturn I and V is the Launch Escape System (LES). This unimposing tube (**4**) is actually a fully armed solid-propellant rocket with the power of 4000 automobiles. Its purpose is to blast the Command Module to safety should a dangerous situation develop on the pad or during launch (up through second stage burnout). Hopefully, the LES's services will never be needed during Apollo. The Saturn V is now complete and stands 363 feet high, equivalent to a 34-story building. Fully loaded with propellants, it will weigh 6,400,000 pounds, of which only 500,000 pounds is the vehicle itself and its payload. Thus 92% of the Saturn V's launch weight is propellants; it is an immense flying fuel tank. Each of the five F-1 engines gulps propellant at the staggering rate of 3 *tons* per second; during launch, then, the Saturn becomes lighter by *fifteen tons every second,* but so huge is the first stage that it will continue to feed propellant for 2½ minutes before it stops firing. The Saturn V is a monument to the capability of our country.

3

4

5

APOLLO 4

It is the morning of November 9, 1967. Before the eyes of the world this first, unmanned flight of the Saturn V is about to determine the integrity of the lunar launch vehicle, the wisdom of "all-up" testing, and perhaps even the fate of the entire Apollo program. Hundreds of reporters are assembled at the press site. They are about to witness the culmination of six years of intense preparation in the greatest research and development effort ever undertaken. As everyone well knows, the "all-up" flight is a risk deliberately taken to keep intact the ultimate goal of putting a man on the moon by 1970. The appalling schedule slippages in the production and qualification of major assemblies, along with spiralling costs, have made painfully evident the ambitiousness of the program. Perhaps the schedule is too tight. The memory of the tragic fire in the Command Module that took the lives of astronauts Grissom, Chaffee and White nine months ago is only too clear. Disaffection with the program, made plain by slashing cutbacks in NASA funding, have now forced a strategic decision. No longer can merely one major component be tested in a flight. So far, only the third stage and instrument panel are flight-proven. Apollo 4 therefore is the first flight for the F-1 engines, the first stage assembly, and the second stage. But these components will do more than merely fly—they have a mission. They will push the Command Module far out in space, turn it around, and then thrust it back at the earth at 25,000 miles per hour, simulating a return from the moon. The CM's heat shield must survive re-entry from this speed, far greater than the 17,500-mph speed of orbital vehicles.

NASA
APOLLO
NASA

APOLLO 4
Pre-launch

After manufacture and testing, the major components of the Saturn V make their way from across the country to the Vehicle Assembly Building (VAB) at the Kennedy Space Center for assembly and checkout. The first stage, or S-IC, F-1 engines arrive from Rocketdyne in California to the Marshall Space Flight Center in Huntsville, Alabama. There the 138-foot first stage is fabricated (**1,2**), checked out, static fired, checked out again, and shipped by barge to Cape Kennedy. The 81-foot second, or S-II, stage, complete with its Rocketdyne J-2 engines, is sent by barge from North American in California to the Mississippi Test Center. There it undergoes static test firing and then is barged to the Kennedy Space Center. The third, or S-IVB, stage is fabricated by Douglas Aircraft in Huntington Beach, California, then flown across the country in a specially designed aircraft called the "Super Guppy." Here we see it in the VAB (**3**). The instrument unit arrives from IBM; the Command/Service module is flown in from

1

2

3

4

5

the west coast; and the Grumman-built dummy, or boilerplate, Lunar Module test article arrives by special aircraft from Bethpage, New York. Replacing the astronauts in the Command Module is an electronic autopilot. Otherwise the Apollo 4 is exactly what the moon mission vehicle will be. After assembly and checkout on the mobile launcher, a huge crawler/transporter moves under it, picks it up, and takes it out of the VAB (**4**). Moving one mile per hour down the 3½-mile specially-built roadway, it takes 10 hours to reach Pad A of Complex 39 (**5**). The missile sits atop and covers a hole in the mobile launcher which allows part of the engine exhaust to be deflected into the flame trench. The Launch/Umbilical Tower is an integral part of the mobile launcher and must stay next to the missile at lift-off. The crawler/transporter then moves the tower-like Mobile Service Structure to the Saturn V. The pneumatic, propellant and oxidizer lines are then connected to the launch vehicle and the countdown demonstration test, or CDDT, begins. This is a full dress rehearsal. It is the first time for the missile, the launch facilities and the crew. Instead of taking 3½ days however, it extends over a three-week period. Trouble arises constantly from a wide variety of equipment. Numerous components are replaced. Fuel and oxidizer tanks are twice unloaded and the countdown moved back. On October 30th, the actual 104-hour countdown begins. This time only two holds are called. It goes so smoothly that a one-hour stop is called, to ensure a day launch.

APOLLO 4
Launch

On November 9, 1967 at exactly 7:00 AM EST the gigantic F-1 engines of the Apollo 4 ignite (**1**). The roar is indescribable. To the reporters and observers at the press site (**2**) 3½ miles away, it is thunder. Shock waves shake the trailers and rattle furniture, and inside the newly constructed Launch Control Center (**3**) everything is covered with plaster dust. Only nuclear test explosions have ever produced stronger man-made seismological readings. After 6 seconds, 90 percent of the 7,500,000-pound maximum thrust is built up. The holddown arms of the launcher deck are released and Apollo 4 slowly rises. In the next 10 crucial seconds, the Saturn V successfully clears the umbilical tower. Several small fires break out on the tower, but damage is minimal. As the instrument section takes over and automatically steers the vehicle eastward over the Atlantic (**4**), one can see again the awesome power of the first stage engines as they produce a luminescent exhaust three times the length of the vehicle itself. Vehicle control is now shifted to the Manned Spacecraft Center at Houston. One minute after lift off, the velocity builds up to 2500 feet per second. After nearly two minutes it is 15 miles away. The inboard engine cuts out 30 seconds later, followed in sixteen seconds by the outboard engines. Altogether, the first stage burns 150.8 seconds. Apollo 4 is now 39 miles up and travelling over 6000 miles per hour. So far it is a textbook flight, but the next series of events is so crucial that special effort has been taken to film them.

2

3

4

APOLLO 4 Mission

A pod capsule camera mounted on the second stage catches the action. With only a three-foot clearance, the interstage adapter ring *must* slip past the second-stage motors without touching them. A solid-fuel rocket on the ring is fired (**2,3**). Then the skirt separates and falls away (**4,5**). As the second stage ignites and telemetry signals clearly indicate the success shown here, the flight controllers and observers at Houston cheer excitedly. The cameras are now ejected, parachuted down about 300 miles from Cape Kennedy, and are immediately recovered by Air Force pararescue personnel. The second stage boosts the vehicle to near orbit and then separates. Five seconds later the third, or S-IVB, stage takes over, increases speed to 17,500 miles per hour, puts the Apollo 4 into orbit (**1**), shuts down and remains attached to the vehicle. While orbiting the earth, Mission Control at the MSC in Houston constantly monitors the functions of various spacecraft systems. For almost all in-flight functions, the guidance and navigation

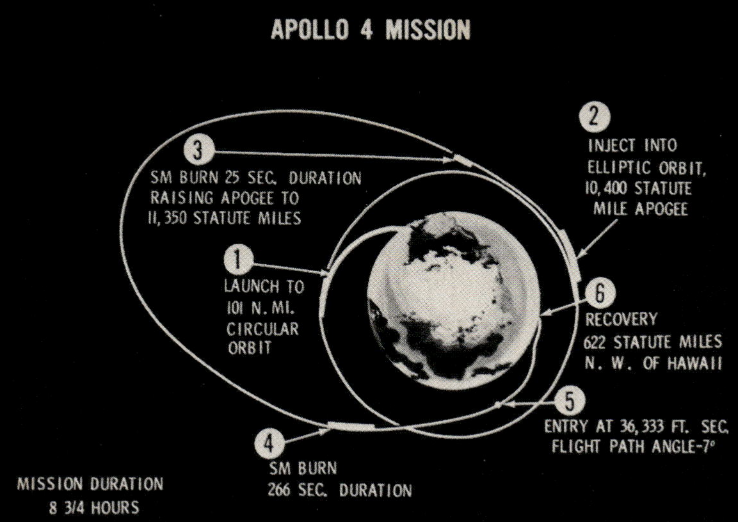

1

2

4

autopilot in the spacecraft controls the vehicle in orbit. However, Mission Control can send commands in case of failure. During the second revolution, the third stage J-2 engine is programmed to restart, but telemetry signals from the vehicle show low hydrogen tank pressure. Has the three-hour ride in freezing space caused a crucial valve malfunction? Other indicators show, however, a proper mission profile. Mission controllers hope that the engine will restart. At three hours 11 minutes after lift-off, the engine does fire and produces full thrust for the scheduled five minutes. The spacecraft now starts on the long elliptical orbit. After third stage burnout and separation, the Service Module engine ignites to drive the Apollo 4 to an apogee of 11,234 miles—far out in space, but a little below the planned peak. Now over three hours into the planned 8¾ hour mission, everything is going beautifully. The next five hours pass uneventfully as the spacecraft follows its long, looping path around the earth. The next sequence of events is the final critical test of the flight. In a moon flight test, the Service Module's service propulsion system (SPS) must restart 36 times after extended cold soaks. How will it respond in this real flight? Now the second firing is to drive the space vehicle earthward, duplicating the 25,000-mph speed a space ship returning from the moon will have. Five helicopters take off from the carrier *Bennington*. At 8 hours, 10 minutes into the mission, the SPS restarts and burns until the Houston cuts it..

3

5

APOLLO 4
Splashdown

The spacecraft now rushes toward earth at 25,000 miles per hour. On another signal from Houston, the Service Module seperates from the Command Module. A final radio command directs the autopilot to turn the Command Module around so that the heat shield faces the earth. With atmospheric friction, the spacecraft is subject to 5000 degrees F. as expected and a peak deceleration load of 8 gravities—somewhat above nominal. The descent trajectory is well established and the projection is that it will land very close to the target area 275 miles northeast of Midway Island. In fact the helicopter recovery force already aloft actually spots the spacecraft while descending. Splashdown (**1**) occurs only six miles from the planned spot, 8 hours and 38 minutes after lift off. Navy frogmen immediately jump in, swim to the craft and attach the flotation collar (**2,3,4**). Later, the *Bennington* arrives and picks up the Command Module (**5**). Post-flight examination shows that the heat shield succesfully withstood re-entry heat of

1

2

3

more than than 5000 degrees. Furthermore, the temperature within the Apollo spacecraft never climbed over 70 degrees. Total mission performance has been nearly perfect. This is an astounding achievement considering its complexity. For example, the numerous attitude control engines as well as main stage engines, 95 in all, functioned perfectly. The implications of this resounding success are both profound and far-reaching. The recently acquired sophistication in simulations, environmental testing, quality control techniques, reliability engineering, and scale model testing has proven the validity of "all-up" flight testing. Also contributing mightily is the high degree of automation in the checkout and countdown procedures. Furthermore, the expectation for success is now more like that of aircraft development, rather than the sometimes embarrassing hit-or-miss experience of some earlier research. This maturity has come none too soon. Whereas public support for space programs had been considerable, attention is now focused upon other pressing political issues. Had there been a major failure, if the Saturn V did not so convincingly prove itself, the chances of putting men on the moon by 1970 would be dim. Instead, however, we now have a winner. There is no question that sufficient support for Project Apollo will be forthcoming. The nation and the world alike now know that the moon is truly within our grasp. Next is an unmanned flight test of the vital Lunar Module, then we will be ready to put men in our Apollo modules.

4

5

Project Apollo represents the largest peacetime enterprise in the history of American business. The products and services required so that men may get to, live in, and return from deep space are a varigated lot, as are the companies which furnish them. Since much of what is needed must be invented, when NASA goes shopping it follows a well-defined system. This contract system results in the design, production, testing and use of the hardware and back-up software which together form Apollo. The uniqueness of it demands that extraordinary care be exercised at every step of the process. A requirement, often in the form of a design concept, is formulated by NASA for a particular item or group of items. The agency then circulates specific design requirements among qualified industrial producers, inviting them to submit working designs along with cost estimates for their particular proposal's production. Since it is a competitive system, several producers usually submit bids for the same contract. NASA evaluates each re-

CONTRACTORS & SUB-CONTRACTORS

A clearer view of America's technological arsenal

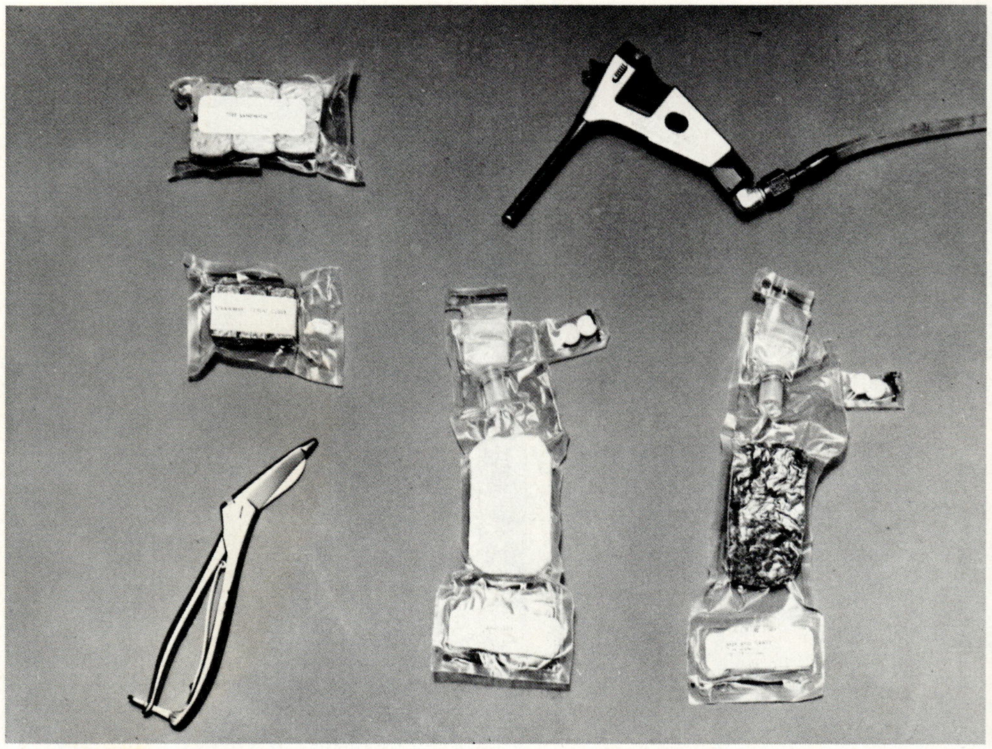

1

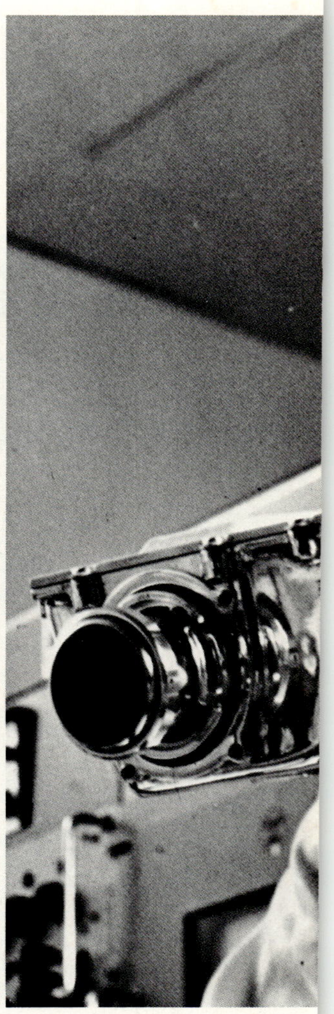

2

3

sponse, taking into consideration not only cost but also such factors as the company's past technical and managerial performance as well as its individual design's ability to integrate with the entire unit of which it will form a part. This criterion of compatability is vitally important in a technological effort so vast and complex as the Apollo spacecraft and support systems. Thus each potential supplier challenged to meet NASA's requirements must display much advance interest and competence if it hopes to get the contract. There are two classes of contractors. Prime contractors are charged with supplying a major piece of hardware. A sub-contractor is asked to supply a component, in which it specializes, to the prime contractor. For instance, Grumman Aerospace Corp., prime contractor for the Lunar Module, has more than 100 firms in specialty fields supplying components for the LM. RCA Whirlpool, known for their home appliance products, developed the prepackaged food system for Gemini and Apollo (**1**). Special cutters open the packs, which are then rehydrated with a sophisticated squirt gun. Another famed American appliance-maker, Westinghouse, built this compact TV camera (**2**), though a Swedish company, Hasselblad, has been the NASA still-photography supplier since Mercury (**3**), and has designed a special four-camera mount (**4**) for Apollo. International Latex Corp. constructs the Apollo pressure suits—influenced by the Mercury suit built by B.F. Goodrich, and the Gemini suit formulated by the David Clark Co.(**5**)

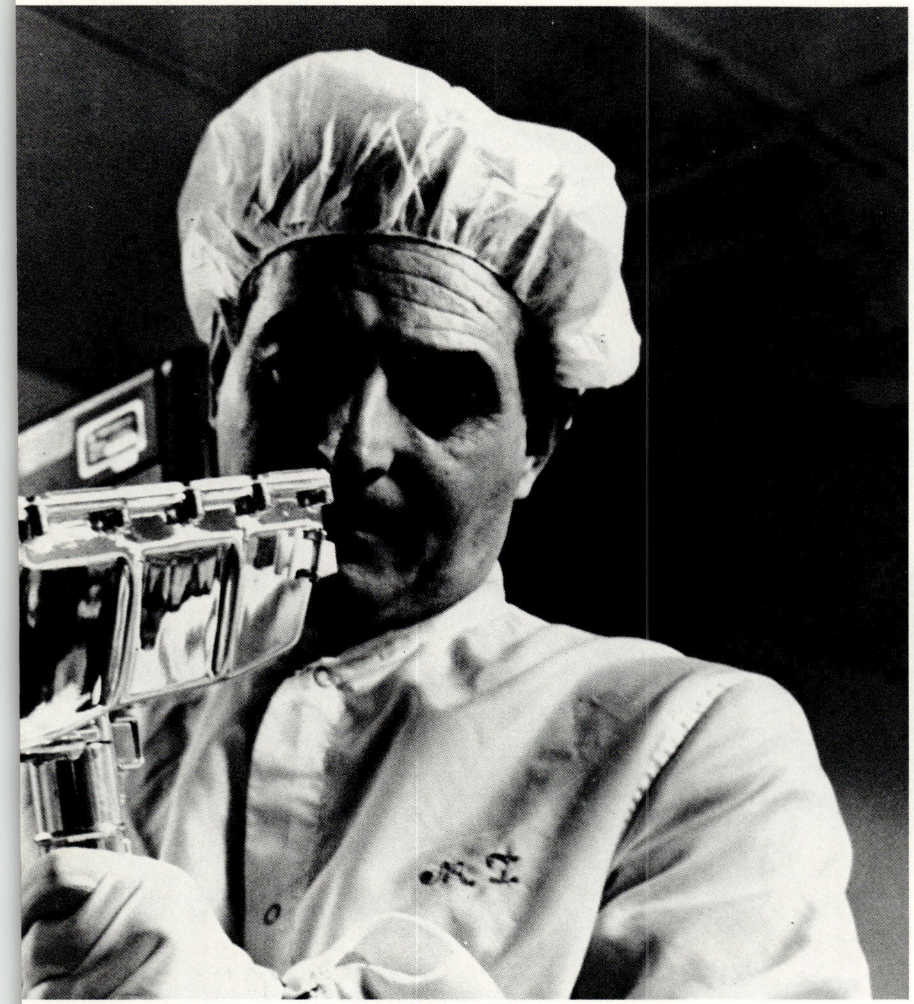

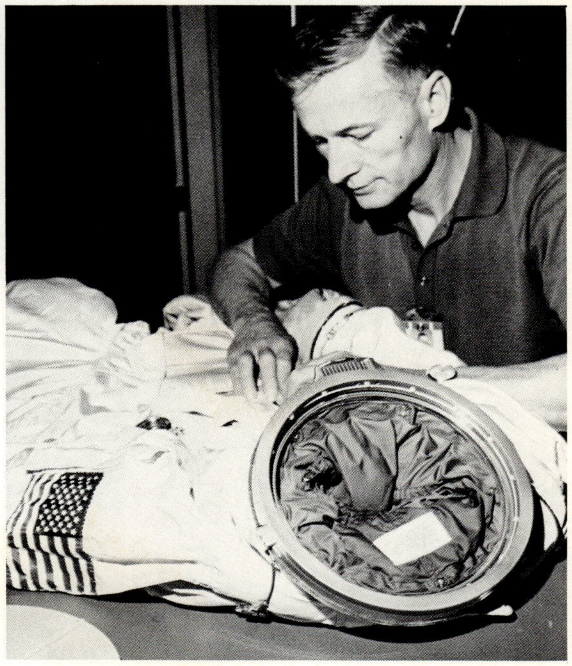

5

4

CONTRACTORS & SUB-CONTRACTORS

Contracting on a nationwide basis, NASA faced numerous logistics problems in shipping space hardware. Smaller components could be handled by conventional methods, but outsize and delicate items were another thing. Aero Spacelines came into being in 1961 to solve this problem. Their solution was to modify a Boeing C-97 Stratocruiser (**1**), quickly FAA-certified and dubbed the *Pregnant Guppy*. Its immense ariframe separates (**2**) allowing even large components like Saturn's S-IV stage to be transported. The Santa Barbara-based California company has been NASA's prime civil air carrier since 1963 and has airlifted Agena, Apollo, Centaur, Gemini, Saturn, Surveyor and Titan components. A larger version (the *Super Guppy*) was constructed (**3**) and flew in August 1965. It featured a swing-away nose for cargo loading and four Pratt & Whitney R-4360 radial engines with Hamilton-Standard propellers for power. The Pregnant Guppy has a cruise speed of 225 mph and payload capacity of 29,000 pounds. The Super

1

PHOTO COURTESY AERO SPACELINES

2

Guppy can carry 41,000 pounds and cruise at 300 mph. The free-enterprise system again advances a unique solution to a difficult problem created by the space age. To fill NASA's needs, some companies, like Hasselblad of Sweden, modified what they made. Others like Aero Spacelines, Inc., anticipated a need and organized to fill it before even seeing a government contract. The majority, however, were companies who were well established in government contract work or had impeccable reputations for consumer products. Bendix Corp., of Ann Arbor, Michigan, combines both. In March, 1966, Bendix won a prime contract from NASA, over Space General Corp. and TRW, for development and manufacture of portable lunar scientific stations to be carried with man to the moon. NASA's Lunar Exploration and Sciences Conferences and the National Academy of Sciences Space Board set down the requirements in 1965. Bendix developed 15 experiments in conjunction with researchers, four with components supplied by the government. Reducing the packages to stowable size and reasonable weight called on modern methods of miniaturization (**4**). Each package measures about 24x27x21 inches in size. But Bendix Aerospace's most difficult problem is making certain their packages can survive the extremes of lunar temperatures. A special chamber (**5**) was built to simulate the +250 to -300 degrees F temperatures the stations will experience. Bendix developed a plutonium fuel that will keep working for years.

3

PHOTO COURTESY AERO SPACELINES

4

PHOTO COURTESY BENDIX

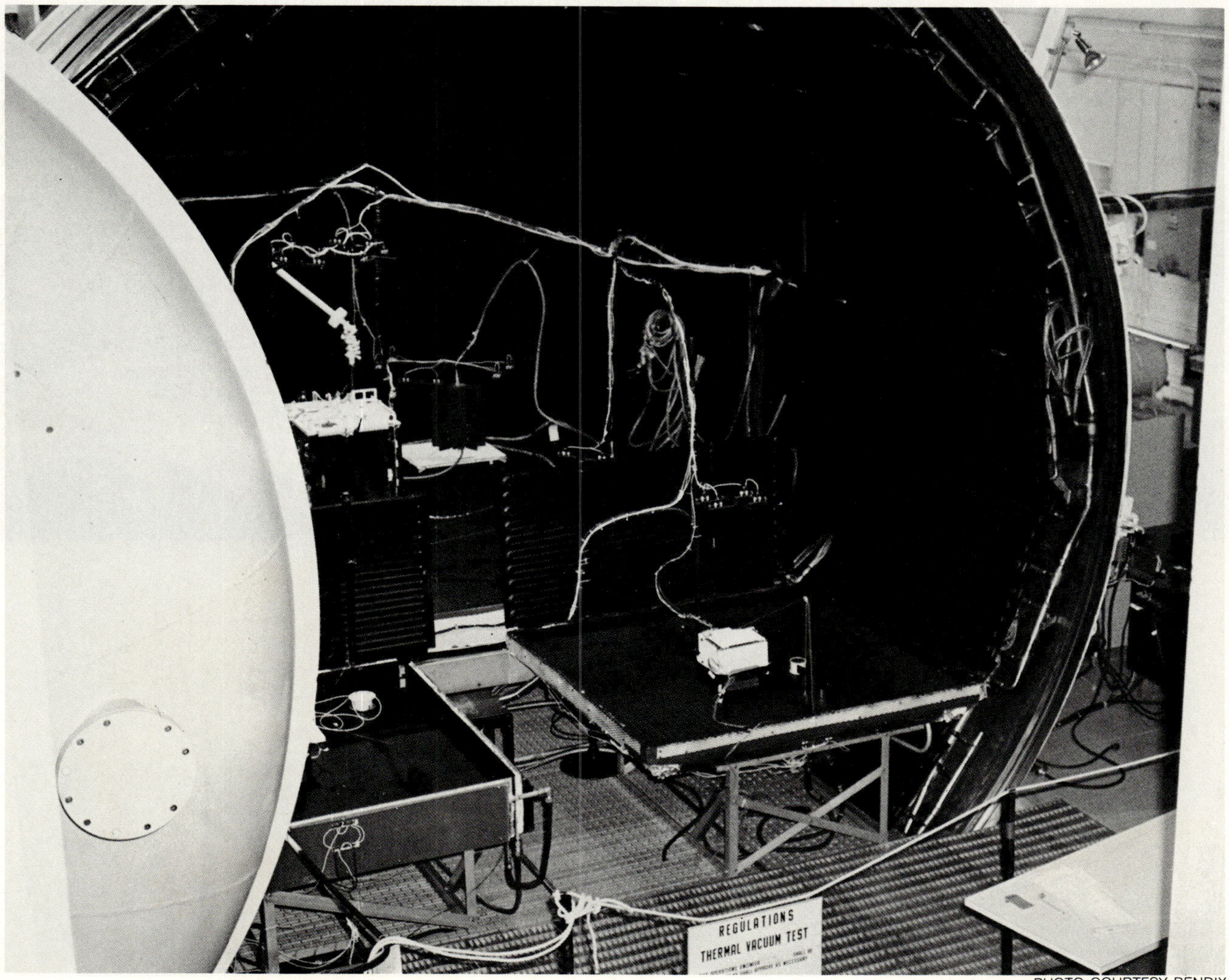

5

PHOTO COURTESY BENDIX

CONTRACTORS & SUB-CONTRACTORS

Propulsion—that is what is needed to take us to the moon, and some years ago, it was an area where we stood in second place to the Soviet Union. But the Rocketdyne Division of North American Aviation stood ready to the task when initially approached by the USAF. The Air Force contracted with Rocketdyne on June 23, 1958 for preliminary design of a rocket engine that would produce at least one million pounds of thrust. The project was subsequently transferred to NASA on December 12, 1958. Designated F-1, the prototype was fired on February 10, 1961, producing the requisite thrust. Rocketdyne also undertook the design and development of the J-2 and H-1 engines for Saturn, placing the company in the forefront in rocket engine design. Proudest achievement of the Canoga Park, California-based company, the F-1 (**1**) is the largest thrust engine in the world. One of the most widely debated facets of the lunar landing program was: which method was best? Confident of our technology, several proposals were batted

1

back and forth for a number of years. Grumman Aircraft Corp. began manned spacecraft studies in 1958, and in 1960, submitted a preliminary study of Project Apollo to NASA. Prevalent at that time was the earth-orbit rendezvous flight plan. A combined ITT/Grumman study was not successful in winning the NASA contract. By mid-1962, Grumman was convinced that the lunar-orbit rendezvous method was the most realistic, and began studying the LOR concept. Grumman, with RCA as principal subcontractor, submitted its findings to NASA in September, 1962. NASA, committed only since July to the LOR concept, was impressed with the Grumman/RCA report. On November 7, 1962, Grumman was awarded the Lunar Module contract (**2, 3**). Some 8000 people are involved in the LM program. On November 28, 1961, NASA awarded the design and development contract for the Apollo spacecraft to North American Aviation's Space Division. Shortly thereafter, the prime subcontractors were announced. They included Collins Radio, telecommunications; Garrett Corporation's AiResearch Division, environmental systems; Honeywell, stabilization and control; Northrop's Ventura Division, parachute earth landing system. These and other organizations across the country have united in a common goal, to make the Apollo spacecraft (**4**) a reality. It is shown here with Launch Escape System, built by Lockheed, in place. North American's plant in Downey, Calif. (**5**) is the spacecraft's place of birth.

2

PHOTO COURTESY GRUMMAN

4

3

PHOTO COURTESY GRUMMAN

5

PHOTO COURTESY ROCKWELL INTERNATIONAL

CONTRACTORS & SUB-CONTRACTORS

Also among the giants of American industry which answered JFK's call is Boeing Aerospace Company of Seattle, Washington. Perhaps more passengers have flown on Boeing jetliners than on any other manufacturer's. Hopefully, they'll become as well-known for their S-IC and LRV as they are for their 700-series jetliners. To Boeing goes the contract for constructing the 33-foot diameter S-IC first stage for the Saturn V. Though not so large as the company's bigger planes, this top for a fuel tank (**1**) still presents enormous engineering problems. The weld joining this piece with the body of the tank, for instance, is over 100 feet long; it requires absolute precision in fabrication and measurement, as well as revision of some current welding methods. As is the procedure with other contractors, Boeing personnel stay with the company's products from factory through assembly in the Cape's VAB (**2**). While Boeing is building big, another huge corporation, IBM is thinking small. Shown here (**3**), along with some of its com-

1

COURTESY BOEING C

2

COURTESY BOEING CO.

3

COURTESY I

ponents, is the computer International Business Machines' Federal Systems Division built for the Gemini spacecraft. A dominant force in the design and manufacture of computers (so much so that the company's name and its field are almost interchangeable), IBM has developed products for every level of the nation's public and private sectors. Its Federal Systems Division (**4**) located in Huntsville, Alabama, near the Marshall Spaceflight Center, has been awarded a contract for the Saturn V Instrument Unit—focal point of all guidance and telemetry in the Apollo launch vehicle. The other end of Apollo's mission, the Lunar Module's complex descent engine, is a project of TRW of Redondo Beach, California (**5**), one of the nation's largest aerospace firms. Among TRW's more earthbound products are check and credit card verification instruments utilized by many retail stores. The rocket it will build for the LM is the only one designed to operate with a variable output, and the only thrust motor that lowers a vehicle rather than launching it. The massive Crawler/Transporter (**6**) which moves the rocket and its launch tower from VAB to launch site is built by Marion Power Shovel Company of Marion, Ohio. This was the great talent NASA displayed as it searched through America's industrial network for Project Apollo: from coal shovels to computers, no manufacturer was too obscure or too large to be asked to help. If it showed design competence and dedication to the tasks at hand, it got a job and NASA received a product.

4

COURTESY IBM

5

COURTESY TRW

COPYRIGHT 1973 BY CTEIN

CONTRACTORS & SUB-CONTRACTORS

One of the oldest contractors in the "space" business, as well as the government contract business, is (**1**) McDonnell Aircraft Corporation of St. Louis, Missouri. A participant in the space program from the start of Project Mercury as prime contractor for the space capsule, McDonnell continued with the fabrication of the Gemini capsules, too (**2**). Another well-known aerospace firm is Douglas Aircraft Co., which is later to merge with McDonnell. Douglas develops the structure of the S-IVB in its Huntington Beach, California, (**6**) facility. Thus McDonnell Douglas, through one of its divisions or another, has contributed its expertise to every segment of the manned spaceflight program; its employees can be justifiably proud of the parts their companies, individually and collectively, play in the national effort. Douglas was awarded the contract for the S-IVB stage in December 1961 under the direction of the Manned Space Flight Center. The contract, totalling $141.1 million, calls for design, development and construction, plus

1 COURTESY MCDONNELL DOUGLAS

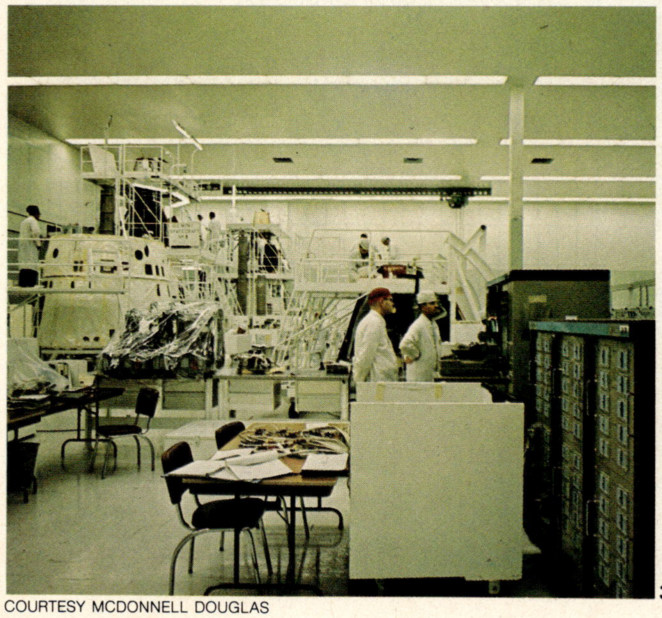

2 COURTESY MCDONNELL DOUGLAS

3 COURTFSY NORTHROP

testing, of eleven S-IVBs. Three units are for ground test, two for unpowered flight, and six for space launch; a completed version is shown (**5**) on its transport trailer before being airlifted to the Cape. Another long-time aerospace contractor involved in the space program from its very beginnings is Northrop Aircraft, based in Hawthorne, California. Northrop's Ventura Division, located in Newbury Park, California (**3**), subcontracted for the Apollo parachute recovery system from North American Aviation—prime contractor for the Apollo spacecraft. Northrop was also among eleven companies invited to submit working bids for the LM and one of nine which completed the necessary studies. Though denied that contract, its previous experience (Northrop developed parachute recovery systems for both Mercury and Gemini) made the company a natural for construction of the final operating sequence in the space flight of Apollo (**4**). These, of course, are but a few of the thousands of companies which responded to NASA's requests. Together they make the food and clothing and the spacecraft; they design the rockets and the engines, move them from disparate points of assembly to launch site, guide them out into the darkness of space and set them on the moon's bright surface; they show us earthrise and lunarscape, and bring us pieces of another world. It is no idle political boast to say a nation will land on the moon, for by the hundreds of thousands Americans worked this technological miracle for all the world to see and marvel.

5 COURTESY MCDONNELL DOUGLAS

6 COURTESY MCDONNELL DOUGLAS

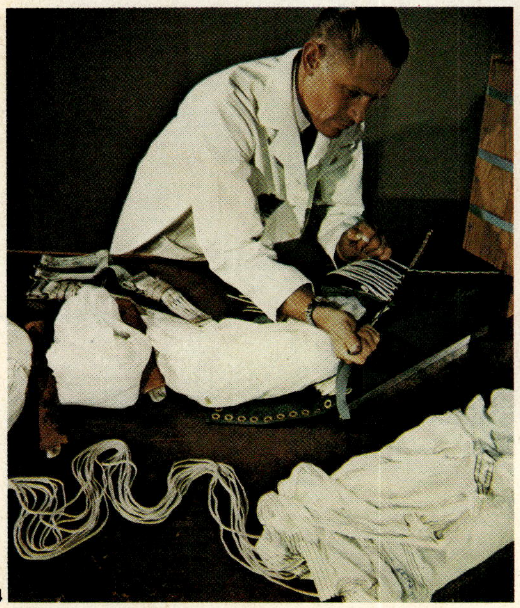

4 COURTESY NORTHROP

APOLLO 5

The date is January 22, 1968, the place is the Kennedy Space Center in Florida where the stage is set for the launching of Apollo 5. Excitement has not yet died down from the magnificent performance of Apollo 4 and the incredible Saturn V shot of November 1967. There seems to be less interest on the part of non-involved people to the pending flight of Apollo 5. The attitude of the news media and general public seems to be, "Ho hum; after an act like the Apollo 4's Saturn V launch vehicle, what can a Saturn-1B do for an encore?" Granted, the thunderous 7.5-million pound thrust displayed by the first Saturn V astounded the world and reached far more ears than the reverberations of those F-1 engines could. It proved that America had the power necessary for a lunar journey. Apollo 5, using the Saturn 1B launch vehicle that was to put Apollo 1 into orbit (AS-204), will be carrying a passenger! A strange looking craft known as the Lunar Excursion Module (LEM). Apollo 5's "act" will be the first orbital test of the two-stage spacecraft designed to carry astronauts from the Command Module (CM) to the lunar surface and back to the orbiting CM. During an actual manned lunar flight, the Lunar Excursion Module's manned upper stage would be flown to the lunar surface by the power of the descent (lower) stage. The manned stage, also equipped with a rocket engine would then separate from the lower stage to take the astronauts back to CM rendezvous. The Saturn V can provide the power, but the LEM has to provide the means for man to land, then return. Testing the LEM (shortened to LM for benefit of the press) is Apollo 5's mission and that's some encore!

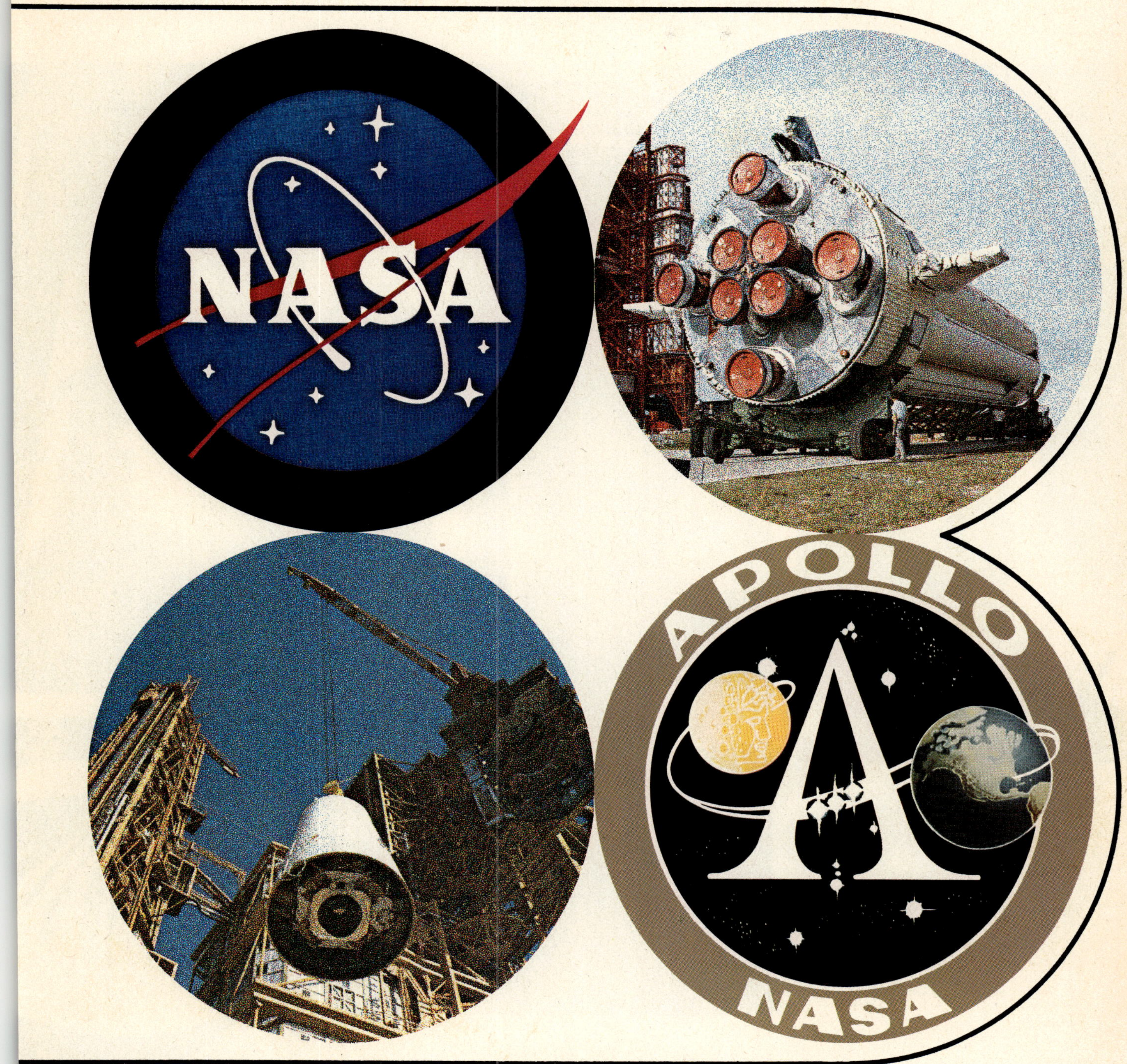
NASA
APOLLO
NASA

APOLLO 5
Preparation

The Saturn-1B launch vehicle, which will eventually boost a manned Apollo spacecraft into orbit, is constructed with eight Redstone (H-1) engines arranged in a cluster, so that the 188,000-pound thrust from each can be combined by firing all eight engines at the same time to boost the spacecraft into orbit. On this flight, the Lunar Module is going along for the ride. But before the spacecraft goes anywhere, the entire "package" must be assembled. The different pieces of the launch vehicle, Spacecraft Lunar-Module Adapter and Lunar Module (LM) are built independently, rather than as one unit. Each stage is then encased in a sheet metal covering known as its skin, and all the stages are fitted together to form one complete rocket. The first stage of the Saturn-1B launch vehicle (**1**) stands inside its service structure, which enables technicians to work on it. The first stage, with its clustered H-1 engines, is built to develop a thrust of 1.5 million pounds. The length of the first stage is 80 feet, and the diameter is 21½ feet. In

1

2

3

4

this view (**2**), the Saturn 1B first stage is being hoisted into place on the launch pad at Complex 37. The second stage, the now familiar S-IVB stage with its interstage coupling ring, goes up, up, up (**3**). Then it is lowered into position atop the first stage and the launch vehicle is once again ready after its 18-month sleep. Basically, the first stage lifts the entire rocket off the ground, and the second stage boosts the spacecraft (in this case, the LM) into orbit. Looking like an enormous dixie cup with a crinkly top, this bird's-eye-view (**4**) shows Apollo 5's second stage suspended from the service structure. Technicians can be seen standing on the various levels, while cars in the background appear no larger than toys. The J-2 engine of the 1B second stage has a thrust of 200,000 pounds. Wrapped in foil like an ungainly and extremely odd TV dinner, is Apollo 5's passenger—the Lunar Module (**5**), being readied for its test flight. A technician can be seen working on top of the Lunar Module, while others are standing by in Complex 37 at the Kennedy Space Center. Apollo 4 has already proven that the Command Service Module (CSM) works perfectly; now comes the first orbital test of the Lunar Module—the vehicle designed to carry men to the lunar surface and back to the waiting Command Module. Since no alternate plan has been designed to get astronauts to and from the lunar surface it is essential that the Grumman-built Lunar Module perform perfectly on this flight. If it doesn't, the timetable will suffer, with more delay-

5

APOLLO 5
Pre-launch

Assembly of the Apollo 5/Saturn-1B continues as launch day nears. No matter how many times assembly of different spacecraft has been accomplished in the past, there is always an air of tension and excitement accompanying each step of the way. Meticulous attention must be paid to every detail during pre-launch operations. The slightest miscalculation on the part of men or machinery could cause delay of the launch at the very least, or, at the very worst, failure and an aborted mission. Failures in the cooling and electrical systems; malfunction of an engine— . . . . almost anything could go wrong. It is the responsibility of all the people involved in putting together the components of the Apollo/Saturn-1B to see to it that everything goes *right*. The Saturn-1B second stage (**1**) is shown being raised in the service structure for mating to the first stage. Gently, gently, the second stage is maneuvered into position (**2**) as it hovers over Saturn's first stage. The huge crane moves slowly along the gantry as the S-1VB

1

2

stage is lined up precisely with the first stage. Machine-milled waffle plates, looking very much like a waffle iron or an old-fashioned toaster grid, can be seen under the crossed girders, used for lifting only, clamped to the rim of the second stage. Technicians peer intently into the "guts" of Saturn's first stage, as the stage is moved carefully into place. Here it is (**3**): the little beauty. It looks like a huge bell, but what we see suspended from the very top of the service structure is the Spacecraft Lunar Module Adapter (SLA), being readied for mating to the second stage of the Saturn 1B launcher. The base of the Lunar Module, to be used as a lunar launching pad to get future astronauts off the moon and back to the Command Module, can be seen to the right of the technician's hard hat. As the Spacecraft Lunar Module Adapter (**4**) is being hoisted for mating with the Saturn launch vehicle, the landing gear of the enclosed Lunar Module shows up clearly. The mating of the SLA is taking place in Complex 37 of the Kennedy Space Center in Florida, prior to the proposed launching of Apollo 5 on January 22, 1968. This particular mission is extremely critical, because it will test the feasibility of using the LM as the vehicle to ferry the astronauts from the Command Module to the lunar surface. If the LM doesn't work as planned, it's back to the drawing board, which will mean a delay of manned lunar landings. The LM, complete except for unneeded landing legs, will be put through its paces by remote control.

3

4

APOLLO 5
Launch

The week leading up to the launch of Apollo 5 has been plagued with bad weather. The skies have been overcast, with intermittent rain; the temperature chilly. Weather conditions have hardly been auspicious for the scheduled January 22nd lift-off, even though launch activities continue on a full scale. For a while it looks as though the Saturn-1B will not take off on time after all. First a power failure blacks out the electronic computer monitoring systems aboard the Saturn launch vehicle and delays the start of the 22 hour countdown. Then there are frustrating failures in the cooling and electrical systems of the ground equipment at Complex 37, delaying the launch nearly four hours. In addition the Saturn rocket has been parked at Cape Kennedy for nearly 18 months, and no one will know whether or not it will work until the actual launch. There are no trial runs: the engines are fired and the rocket either takes off perfectly—or it doesn't. Proper repairs and adjustments are finally made, and countdown begins. Standing 180

1

2

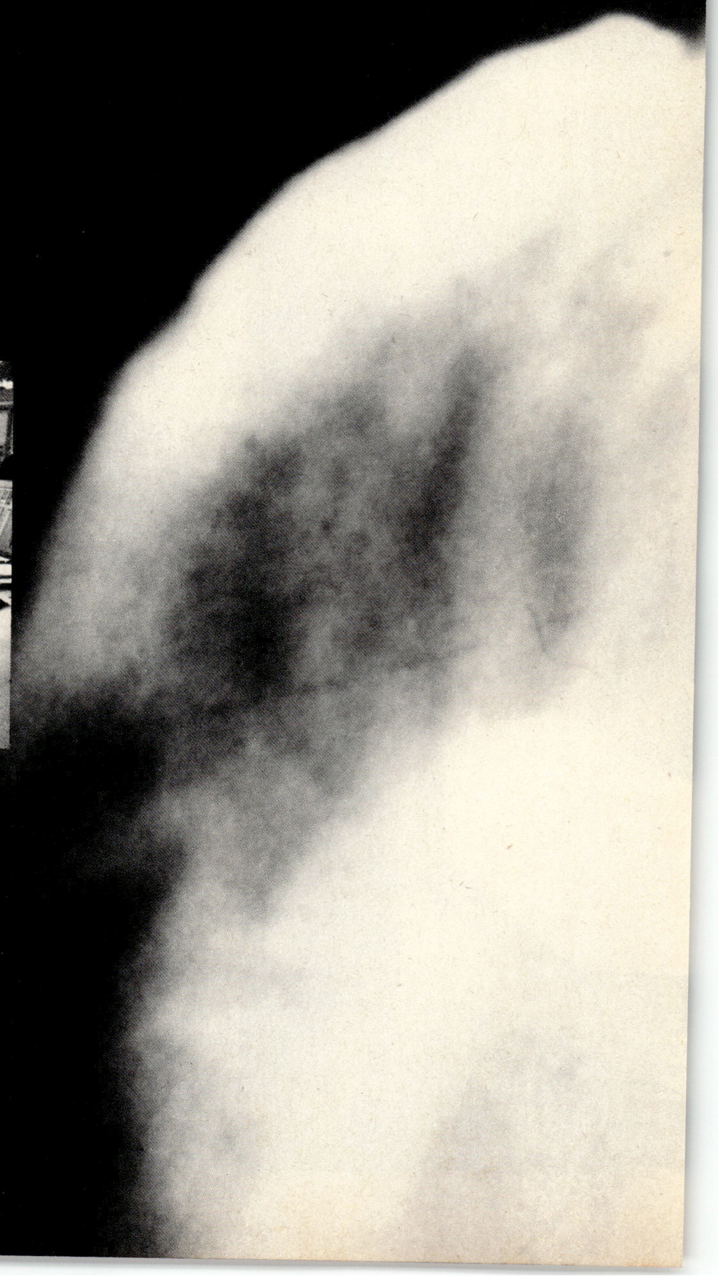

feet high, Apollo 5 pokes its nose into the sky (**1**) waiting on its launch pedestal, and still attached to the launch umbilical tower by the tower's service arms. The Lunar Module, mounted on the Saturn launch vehicle, is neatly tucked away inside its "skin". Because this mission is a test to determine how well the Lunar Module will withstand the rigors of space travel, as well as determining how well it will function, the LM has been placed in position without legs. The 31,700 pound "Bug" won't need its legs on this trip, since it isn't going to land anywhere. Plans call for the LM to be put through its paces and then jettisoned in space, where it will burn up in the atmosphere. Activities in the Launch Control Center firing room (**2**) are in full swing as countdown gets under way. Eyes riveted on television monitors and computers, earphones and headsets attached to anxiously listening heads, the attitude in the firing room is one of total absorption. It's late afternoon in Florida, the sky is turning dark, countdown is completed and Apollo 5 is off! Saturn's eight H-1 engines ignite the sky in a burst of yellow and orange flames (**3**), thrusting the blunt nosecone into the atmosphere. Lift-off is perfect, and within minutes the LM is injected into an oval orbit of 107 by 138 miles. Apparently the 18-month-long "rest" did the A/S-204 launch vehicle, originally slated to lift Apollo 1 into an earth orbit, no great harm. Test conductors can relax a bit and watch the Lunar Module as it goes into its "act" high above the earth.

APOLLO 5
Mission

The stages of the Saturn launch vehicle have dropped off as scheduled, and everything is calm and orderly as technicians watch the computers and TV screens at Mission Control (**1**). The expended second stage of the Saturn (S-IVB) floats serenely off into space (**2**) after separation from the CSM and LM. The plans for testing the Lunar Module call for the LM's reaction control system to be turned on by computer four hours after lift-off, when the vehicle is to be positioned for the first test of the Descent Propulsion System (DPS). The engine is to fire for 26 seconds at 10 percent thrust; throttle up to almost full thrust, and burn at that rate for 12 seconds, simulating a partial descent from lunar orbit. Engineers at Mission Control are utterly dismayed as the descent engine ignites, runs for four seconds and cuts off when the engine reaches 9.5 percent thrust instead of the programmed 10 percent. As it turns out, the engine is fine, but the failure is in the computer. The engine could have been turned

1

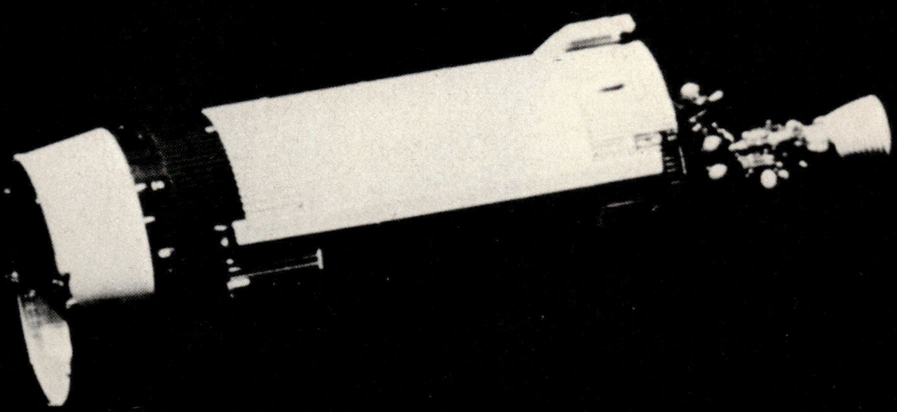

2

back on by the crew, if the LM had been manned. To make sure it doesn't happen again, control is shifted from the primary guidance computer to the Program Reader Assembly (PRA), a much simpler type of autopilot. The PRA simply takes control of the descent engine and operates it according to instructions printed on film. Two subsequent firings of the descent engine perform normally. An artist's rendering (**3**) depicts the LM (this time with legs . . . as it will have on lunar flight) after the first firing of the descent engine. We rely upon the artist again, as he depicts the second and third firings of the descent engine (**4**). As the descent engine is shut down, the LM's Ascent Propulsion System (APS) is ignited and the ascent stage (**5**) is boosted away from the descent stage. The LM's guidance computer has taken over from the PRA during the second ascent burn, and suddenly the LM goes wild! The "Bug" is doing a mad dance in space, and no one knows why. Again, it's the computer, "reading" the LM's performance incorrectly. Corrections made by the autopilot put a stop to the space-dance, and the LM continues merrily and steadily on its way, out of radio range. After five revolutions, the LM comes back into radio range of Hawaii; its ascent stage reenters earth's atmosphere, and on its sixth revolution the little "Bug" disappears once and for all. The future ferry of the astronauts disintegrates as it re-enters the earth's atmosphere, but the mission is a success, and the LM is ready to fly men to the moon.

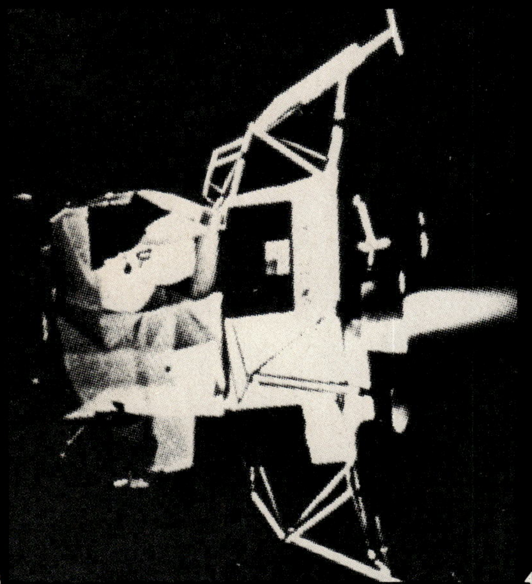

3

4

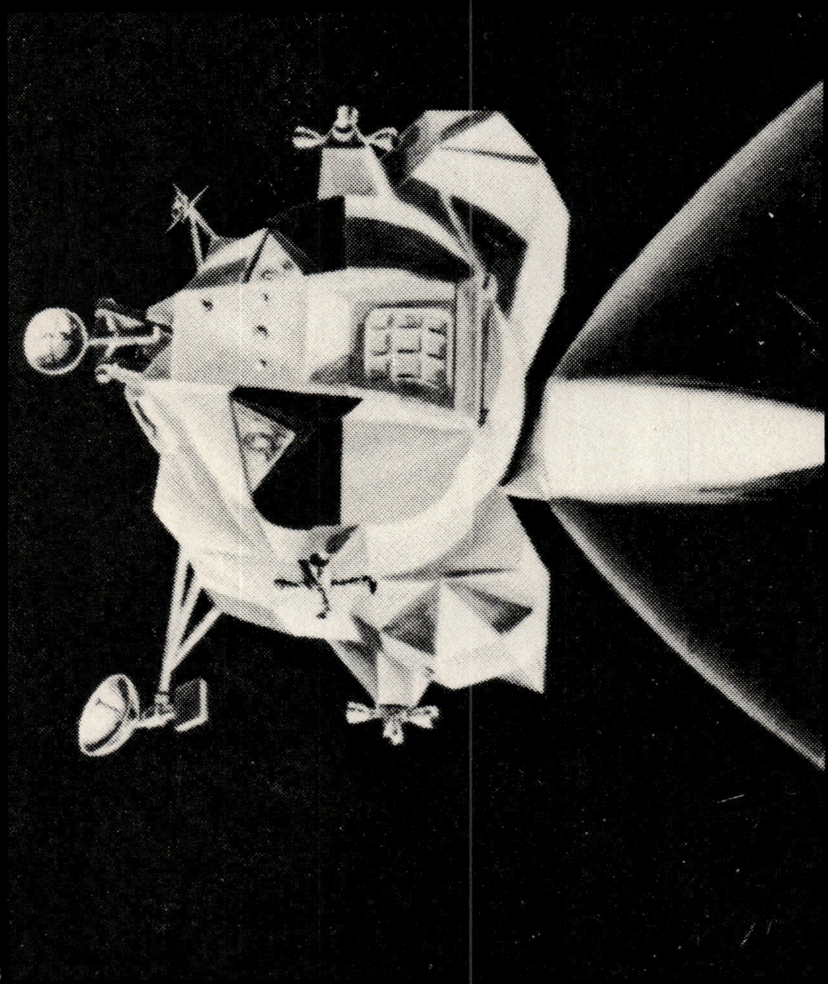

5

"The aeronautical and space activities of the United States shall be conducted so as to contribute . . . to the expansion of human knowledge of phenomena in the atmosphere and space." These words, contained in the National Aeronautics and Space Act, could well be interpreted as the motto of NASA. Since its inception in 1958, the National Aeronautics and Space Administration has dedicated itself to the continuing increase of the entire world's knowledge. To many of us, our only glimpse of NASA comes via television or the newspapers when a significant project is about to begin or has reached completion. Impressive launches at Cape Kennedy and scenes of controllers sitting in front of mysteriously flickering panels at Mission Control in Texas have become familiar sights to all of us—but there is far more to NASA than Cape Kennedy and Mission Control. The activities we see publicly are the fruits of years of research and the products of an organization whose implications are nationwide. And we can be very

NASA'S NETWORK

The people, places and pieces who guide the space program

1 

2 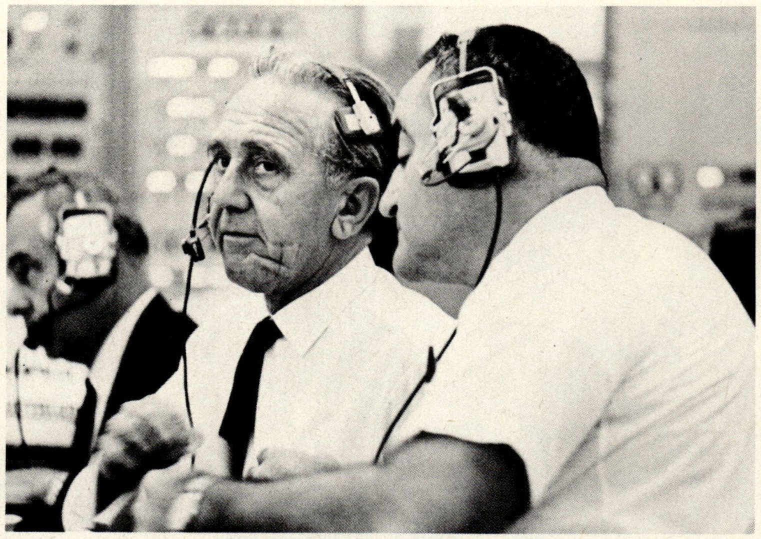

3

4 

5 

proud of it as taxpayers, because the dollars spent towards financing NASA are dedicated to peaceful pursuit of scientific and technological goals that enrichen the lives of all the world's people. NASA is not a concentrated organization, though its purpose is; it is not hidden from public view in the interests of national security, though most of its facilities and people work outside the brilliant limelight of publicity. The seat of NASA operations and management is in the Federal Triangle in Washington, D.C., where George H. Hage (**1**) is the Mission Director and Apollo Program Deputy Director. He is standing in front of a full-scale mockup of the Lunar Module. At Cape Kennedy, the boss is Dr. Kurt H. Debus (**2**), formerly with Dr. von Braun's group at Peenemunde. He is conferring with Dr. Rocco A. Petrone, Director of Launch Operations at Kennedy Space Center. From this site (**3**), Launch Pad 5/6, NASA's first manned flight was launched. Today, it is the U.S. Air Force's Space Museum. In the background is a Redstone launch vehicle and service tower where Alan Shepard was sent on his historic suborbital flight, May 5, 1961. But NASA is not only launch vehicles, service towers, computers and rocket engines. NASA is many people, who by the thousands fill the ranks in order to get the job done. There are engineers and technicians and scientists engaged in many fields, and there are men like this fork lift operator (**4**) and electronics technician (**5**) who silently, but effectively, fill the unglamorous jobs so necessary to our effort.

NASA'S NETWORK

Formerly one element in the Army's Ordnance Missile Command is White Sands Missile Range (**1**), deep in the New Mexico desert. Here, in days past, the Army fired more than 40 captured V-2 rockets under the supervision of a man named Wernher von Braun. When NASA was formed, White Sands was slated to come under civil control, but the Army protested the loss of one of their highly developed test sites. Though still under U.S. Army command, much NASA systems testing has been carried out at White Sands. Marshall Space Flight Center, Huntsville, Alabama (**2**), is the center of NASA operations. Located within the immense Redstone Arsenal complex, it is the training ground for the astronauts, and the testing ground for the Saturn first and second stages. Director of Marshall SFC is Dr. Wernher von Braun (**3**). While still a scientist/engineer under Hitler's Germany, he envisioned the day when rocketry would be used for peaceful ends, and has lived to see an importnat segment develop under his

1

direction. Marshall SFC is NASA's largest facility. All our space projects undergo their birth and development within its confines. One of the "landmarks" at Marshall is the static firing site for the S-II second stage of the Saturn V launch vehicle (**4**). The A-1 test stand is shown here in company with equipment from the Mobile Acoustic Research Laboratory. Effects of noise and vibration are examined which might affect the operation of the vehicle in flight. History in space has been made very rapidly since World War II. Under other circumstances, it might be that an historical branch would be left for the future—with whatever odds and ends might be left over after a project is completed. Such is not the case with NASA, however, which operates a complete museum at Marshall Space Flight Center (**5**). Outdoors is a display of all our major launch hardware, including a Saturn V, its stages separated and lying prone so that the visitor may inspect the 363-foot vehicle closeup. A Lunar Module rests in a circular pit next to the Saturn V. Dominating the vertical aspect is a Saturn I, adjacent to the museum building. Inside the building are numerous displays, many operable by the visitor or guide. Prepackaged space food is available to the visitor, as well as a wealth of educational displays designed to promote public knowledge of the space program. Also represented here is the Army Ballistic Missile Agency. The entire historical display complex is called the Alabama Space and Rocket Center, and is funded by the State of Alabama.

2

3

4

5

NASA'S NETWORK

Residents of New Orleans, Louisiana are familiar with the nearby NASA Michoud fabrication facility (**1**). Here the components that make up the first and second stages of the Saturn V are fabricated prior to their barge trip up the Mississippi River to Huntsville, where they are tested. In closer proximity to Michoud is the Mississippi Test Facility, where NASA will undertake static testing of launch vehicles in the near future. Taking advantage of the immense waterway potential offered by the riverways, fuel supplies for the Mississippi Test Facility is brought in on barges (**2**). At right are two liquid hydrogen barges and at left a pair of liquid oxygen barges. In the background is the static firing station for the Saturn S-II second stage. Far to the west is Edwards AFB, just outside of Mojave, California. Located on the edge of Muroc Dry Lake (**3**), Edwards is well known as the home base for such aircraft as the U-2, YF-12A and X-15. Also stationed there is NASA's Edwards Flight Research Center, established in 1946 under the direction of

1

2

4

NACA, predecessor to NASA. Such famous planes as the X-1 and D-558 flew here; today the hypersonic X-15 operates from Edwards, exploring high-speed flight and re-entry problems in the upper atmosphere. Also undergoing test here are the NASA HL-10 and M-2 wingless "lifting bodies" which should pave the way to maneuverable and reusable space capsules. Also at Edwards is a fascinating, non-flying test facility, the static stand used by Rocketdyne engineers to test the 1.5-million-pound-thrust F-1 Saturn V engines (**4**). Periodically, the desert floor trembles when one of these, the world's most powerful thrust motor, is fired. It was here, on February 10, 1961, that the prototype F-1 engine first yielded a thrust of 1.550 million pounds for a few seconds, indicating that we could indeed launch a manned spaceship to land on the moon. South of San Francisco in Mountain View, California, is Ames Research Center, located at Moffett Field (**5**). Another former NACA site, Ames is one of four NASA research centers. Most prominent are the two huge wind tunnels where various aeronautical and fluid dynamics studies are undertaken, among them the development of VSTOL aircraft. Ames also works on such problems as earth resources, atmospheric entry technology, and planetary missions. The Pioneer series of unmanned planetary probes are built under the direction of Ames. Further research is made at Ames in high-altitude astronautical research, using observatory equipment mounted in high-altitude aircraft.

3

NASA'S NETWORK

All is not strictly space exploration at NASA. At Wallops Island in Virginia, NASA, the Federal Aviation Administration and the USAF have jointly undertaken studies in aircraft and airport safety. A Boeing 727 rolls through an artificially doused runway strip (**1**) to evaluate methods for measuring runway slipperiness. Results of this joint effort resulted in longitudinal grooving of both runways and roads, reducing the dangers of hydroplaning. Also located in Virginia is NASA's Langley Research Center at Hampton (**2**). Langley is among the former NACA stations acquired by NASA when it came into being under the 1958 Act of Congress. Here, many studies are undertaken in connection with space flight, including Lunar Module landing simulators. Technicians (**3**) prepare a Saturn I model in Langley's wind tunnel. The aerodynamic qualities of launch vehicle configurations receive much NASA study long before actually built and fired. Wind tunnel operation is an important facet of NASA research, encompassing not only

1

2

4

launch systems, but also the shape of re-entry systems and advanced aircraft designs as well. In the heart of America is the Lewis Research Center (**4**), adjacent to Cleveland's Hopkins International Airport. Lewis is NASA's chief center for developing power and propulsion systems, and covers 350 acres in Cleveland and 6,000 acres at the Plum Brook Station near Sandusky. Lewis began life as the Aircraft Engine Research Laboratory, and concentrated on powerplant development during World War 2. Today, under NASA, Lewis has pioneered several advanced liquid fuel systems, including liquid hydrogen, valuable due to its light weight. Lewis developed the Atlas-Centaur, Atlas-Agena and Thor-Agena launch systems. A unique installation is Lewis' 510-foot deep Zero Gravity Research Facility (**5**). Here, cylinders filled with liquid are dropped, simulating five seconds' weightlessness. This augments studies of fuel behavior under weightless conditions. The Plum Brook Station houses a nuclear reactor, and is studying the applications of atomic power generation for space systems. Future space missions, with nuclear-generated electrical power, will be able to operate in space for long periods. Nuclear propulsion systems are also under study at Lewis. Atom-propelled thrust engines will be the in-space thrust system of the future. Lewis still undertakes considerable study of air-breathing propulsion systems as well. NASA is many people across the land, studying the problems of the present (**6**) as well as those of the future.

3

5

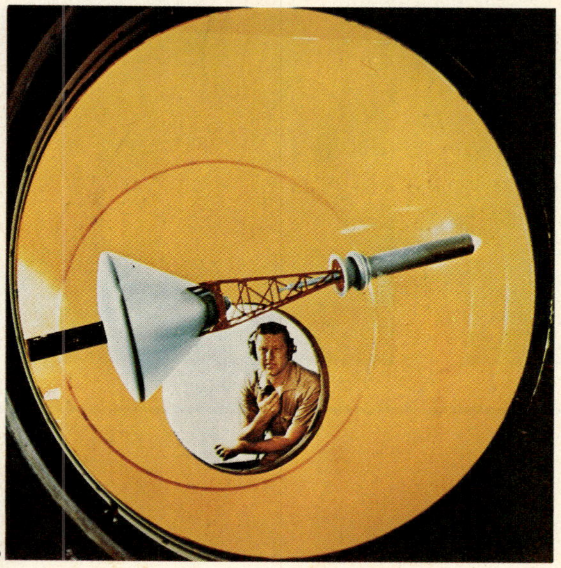

6

APOLLO 6

The last of the unmanned Apollo missions scheduled to be flown, Apollo 6 is a most important flight. This is to be the final in-flight combination test of the Apollo spacecraft and the new Saturn V rocket—America's largest and most powerful launch vehicle. An impressive sight, the Saturn V weighs over six million pounds when fueled, and delivering 7½ million pounds of thrust at lift-off, it develops horsepower equal to that of 453 jet fighters flying at open throttle. Apollo 6 will be an earth-orbital mission, with its prime objective that of qualifying the Saturn V on its second launch as a vehicle for future manned flights. First fired in November 1967, the giant rocket with its cluster of five engines at the base performed almost perfectly at that time. Surprisingly, the launch pad had stood up very well under the blast-off thrust, with only the tail service masts suffering damage when their "clam-shell" like hoods rebounded, allowing engine exhaust to destroy the ground carrier plates. The entire reassembly has been redesigned to prevent a similar problem with the lift-off of Apollo 6 and hopes are now high that this second firing of the Saturn V will confirm its position in America's bid to reach the lunar surface before the Russians do. Secondary objectives of the mission are to test the navigational and guidance system in conjunction with the Service Module's propulsion engine, and another complete test of the Command Module's heat shield. In addition, several propulsion system burns and complete systems testing are scheduled as a part of the Apollo 6 mission. Possible aspects of failure have been considered and alternative plans provided.

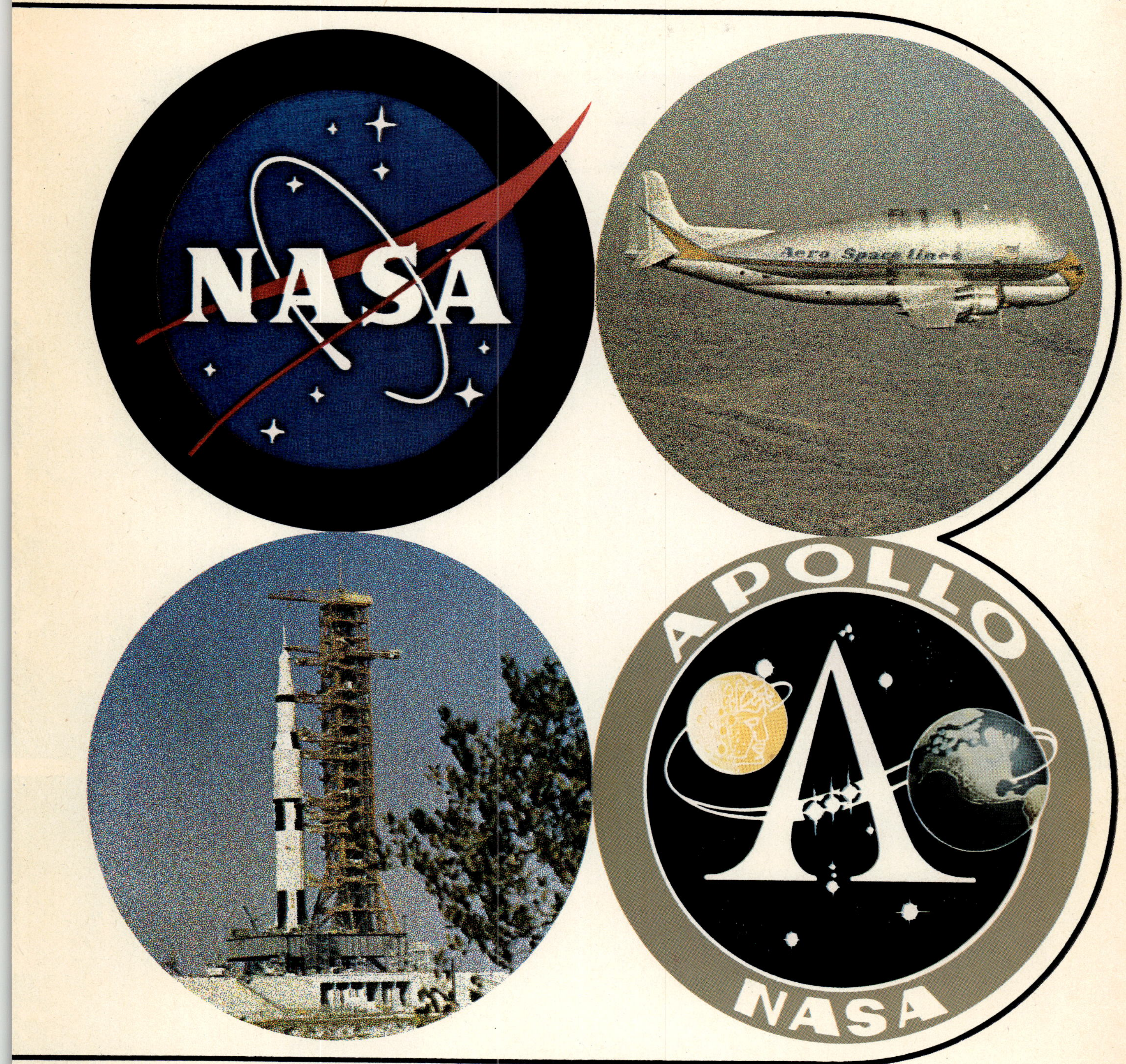
NASA
Aero Spacelines
APOLLO
NASA

APOLLO 6
Assembly

By the end of January, 1968, the S-1C first stage of Saturn V launch vehicle No. 502 arrives by barge at the Cape (**1**). Shortly thereafter it is joined by the S-II second stage and the S-IVB third stage; unlike the preceding two, this last is light enough to be flown in (**2**) by Super Guppy. Then business proceeds as usual for this second launch of Saturn V. The second stage is mated with the first stage (**3**) and to this assembly is added the Douglas S-IVB stage (**5**). A three-foot-deep band describing a circle 21 feet eight inches in diameter, the 4500-pound IU (**4**) serves as the Saturn V brain; it contains electronic equipment for guidance, tracking, and the gathering of environmental and performance data. It also holds the temperature control equipment, batteries and power supplies for all electronic equipment aboard, except equipment in the Apollo modules. All hardware is subjected to a series of rigorous reviews: Design Reviews approve blueprints; Acceptance Reviews assure that, once manufactured, articles in-

1

2

3

deed conform to those accepted designs; Certification Reviews insure that all flight and ground systems can support each manned flight. In this last review vendors, subcontractors, prime contractors, project and program managers, in turn, certify the hardware is reliable and operational; it is completed two to four weeks before actual launch. It should be noted that the Ground Support Equipment (GSE)—the electrical power sources, hydraulic pressure units, and the check out equipment that probes the vehicle's well-being after assembly and through launch—is equally as important as rocket and spacecraft. The vehicle is totally dependant upon the GSE, the complexity of which is indicated by noting that some 60,000 events were monitored during development and manufacture of its electrical and mechanical parts versus 40,000 for the Saturn V's three stages plus IU. Behind these intricate systems is a sophisticated cadre of launch personnel who have worked together since the beginnings of the space program back in the 1950's. In everything from contract negotiations to work in data systems, measurement and instrumentation, computers, inventory management, procurement, price analysis, and data requirements, they have acquired unique experience in military missile systems readily transferable to the space project. Though once they worked directly with vehicles and GSE systems, they now are managers indoctrinating, training, and overseeing the aerospace contractor teams who actually conduct preparations.

5

APOLLO 6
Rollout

Assembly of Apollo 6 continues with the conical CM and cylindrical SM stacked atop the truncated cone of the SLA—all seen here (**1**) swathed in their protective cover cloths, prior to mating with the launch vehicle. Still to be added is the Launch Escape System, a more sophisticated version of the solid fuel rocket and tower assembly not employed since Project Mercury. On February 6, a fully-stacked SA 502 leaves the VAB (**2**) en route to Complex 39's Pad A some three miles away. The great rocket and its Launcher Umbilical Tower (LUT) dwarf the eight-story structure housing launch control as they pass. Reaching the hardstand of Pad A (**3**), SA 502 and LUT are set in place by the Crawler/Transporter which has brought them there. The C/T leaves the site, only to return with the equally sizable MSS (**4**), the arms of which reach out to embrace the rocket. The MSS contains passenger and freight elevators, a power plant, air conditioning equipment, a computer, television and communications

1

2

4

systems, and is designed to provide 360-degree access to upper portions of the vehicle; some seven hours before launch the C/T will remove itself and the MSS, at the time when liquid hydrogen propellant is to be loaded in the upper stages of No. 502. Silver pipes twisting like so many veins, (**5**) the multi-armed LUT continues this work into the night while the rocket waits, steady as its makers' purpose. If only it would remain thus in flight. For a curious series of longitudinal oscillations affected the previous flight of Saturn V. This phenomenon, commonly known as the "pogo effect" because of its up-and-down motions, is not new to liquid-propellant boosters. It occured on Atlas, Thor, and Titan rockets, but in a vehicle of Saturn's size it is more serious. It will not do to have a machine with a human payload shaking itself apart in space or jiggling its delicate components past their tolerance. Basically the problem is one of engine thrust oscillations resulting when vibrations of the rocket's structure couple with vibrations in the propellant feed system. At a certain frequency the one augments the other. The problem has occurred on some of the five engines of Saturn 501, and while the probability of all the engines oscillating in a destructive resonance pattern was considered unlikely, this apparently is what happens in the flight of No. 502. It is subsequently cured by changing the resonance frequency of the liquid oxygen feeder lines—through injection of helium gas—so that their vibrations cannot add to those of the rocket.

3

5

APOLLO 6

Launch

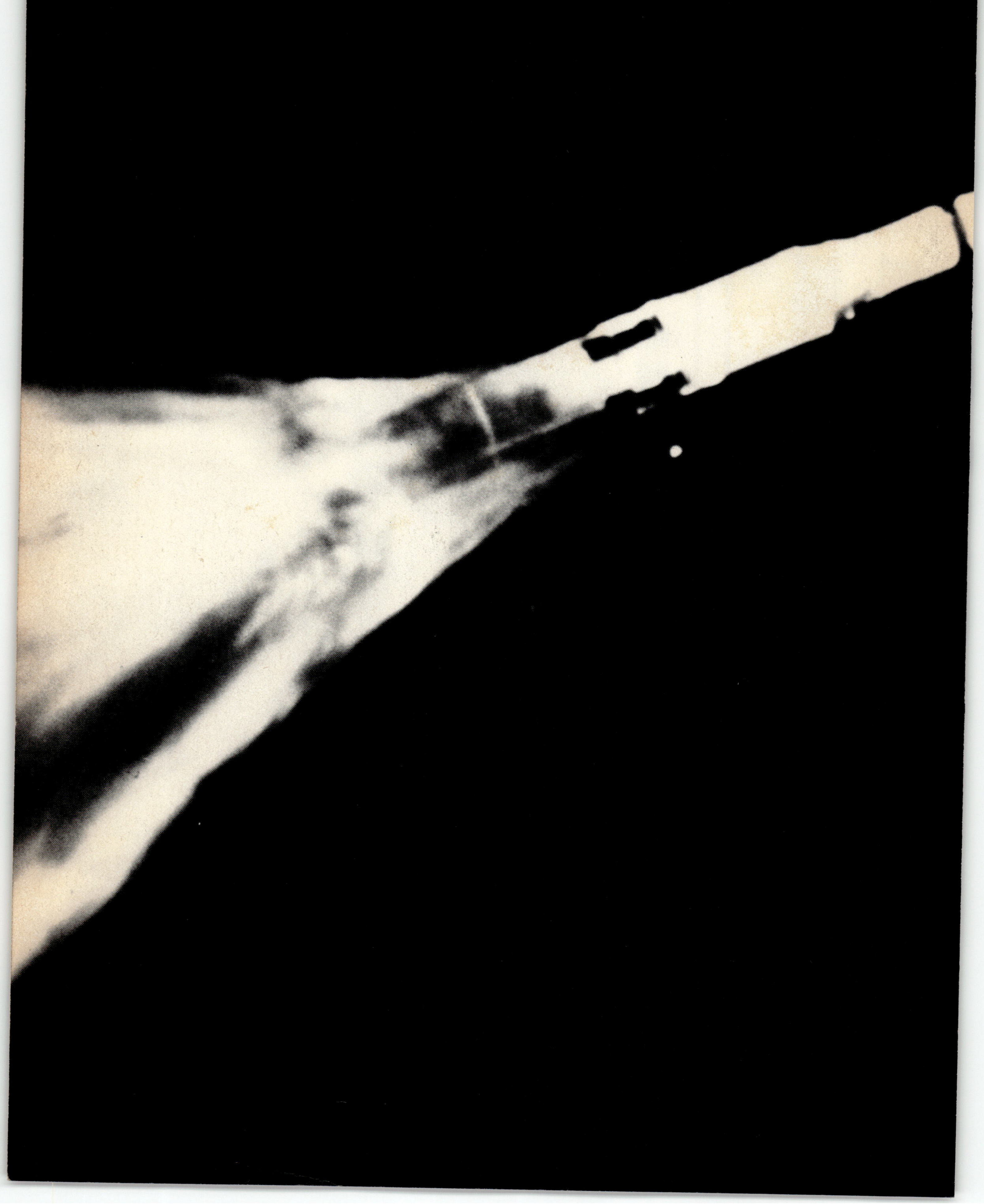

APOLLO 6 Mission

The Apollo 6 countdown ends at 7:00 AM EST on April 4, 1968. Carrying cameras mounted to record the staging operation, the Saturn V blasts off. As soon as their work is done, the camera capsules will be ejected for recovery in the Atlantic off Cape Kennedy. Traveling at an approximate speed of 5400 mph with 148 seconds of burn time elapsed, the five F-1 engines cut out. About 41 miles above the earth, retrorockets on the first stage fire, slowing it to allow separation and "fall-away" from the second stage (**1**). Some four seconds later, when the five engines of the second stage ignite to boost the third stage and spacecraft to a speed of 14,000 mph, the interstage coupling ring also falls away (**2**). As this drops back toward earth, it bursts into flames (**3**) from the friction, exactly what would happen to a spacecraft during re-entry were it not for the heat shield. At this point, problems begin to appear in the remaining stages of the Saturn V. Data telemetered to Mission Control from the rocket indicates that a leak

1

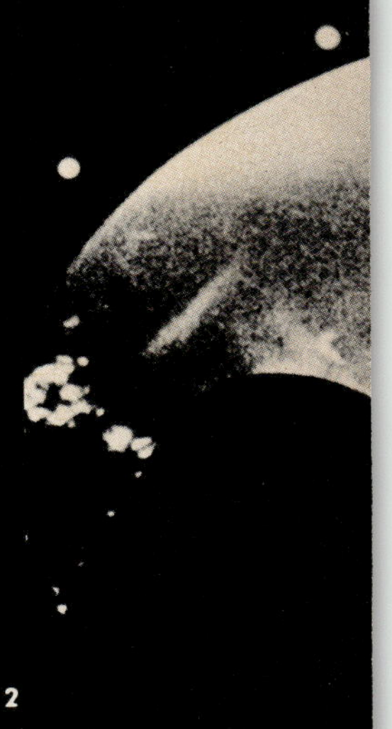

2

4

has developed in a propellent line leading to one of the second stage engines, and shortly after, a loss of power is encountered in one of the J-2 engines. At an altitude of about 120 miles, the second stage is jettisoned and the third stage takes over to put the Apollo spacecraft into an earth orbit. Another leak appears in the propellent igniter line, this time in the third stage engine, causing it to cut off prematurely. All attempts at restarting the shut-down engine fail. These malfunctions in the launch vehicle do not prevent orbiting of the Apollo spacecraft but they do require a change in the flight plan and an alternative mission is selected immediately. As a result of this change, some of the secondary mission objectives must be abandoned, but everything else goes well. The SM propulsion engine is ignited and runs for 7 minutes, 25 seconds, its longest burn to date. The propellent line leaks also serve a useful purpose in pointing up the need to redesign and strengthen them before the Saturn V is used for a manned launch. Passing over Senegal and Mauritania, 1:50 ground elapsed time (GET), at an altitude of 126 nautical miles, a remote camera loaded with high-resolution Aerial Ektachrome film and aimed through the hatch window is triggered to take a photograph (**4**) of the Trarza scrub hills, a large area of elongated fixed dunes. At 3:14 GET, the East Coast of the United States between Savannah and Brunswick, Georgia, is photographed from 100 miles up (**5**); jet contrails can be seen over the Atlantic Ocean.

3

5

APOLLO 6
Splashdown

Once testing is complete, Mission Control rotates the Command Module to position its heat shield for re-entry. Retro-rockets fire, and the Apollo returns to earth, hitting the atmosphere at a speed over 10,000 mph. The heat shield turns white-hot, charring and melting away. Approximately 24,000 feet above its landing spot in the Pacific, the heat shield is jettisoned and a drogue parachute opens to slow the module's speed to 125 mph. At 10,000 feet, the main chute opens, further reducing Apollo's rate of descent to 20 mph for landing at 4:56 PM EST northwest of the Hawaiian Islands. The Navy frogman team on the scene (**1**) rings the capsule with hoist lines in preparation for recovery by the USS *Okinawa.* Once the lines are secured (**2**), the Apollo is hoisted out of the ocean and aboard the carrier (**3**). Apollo 6 is considered a success; tests with the CM went very well, and pinpointing the minor problems with the Saturn V will lead to modification on future launch vehicles. A manned flight is scheduled next.

1

2

3